THE GLOBALIZATION OF SPACE:
FOUCAULT AND HETEROTOPIA

THE GLOBALIZATION OF SPACE: FOUCAULT AND HETEROTOPIA

Edited by

Mariangela Palladino and John Miller

LONDON AND NEW YORK

First published 2015 by Pickering & Chatto (Publishers) Limited

2 Park Square, Milton Park, Abingdon, Oxfordshire OX14 4RN
52 Vanderbilt Avenue, New York, NY 10017

Routledge is an imprint of the Taylor & Francis Group, an informa business

First issued in paperback 2020

Copyright © Taylor & Francis 2015
Copyright © Mariangela Palladino and John Miller 2015

To the best of the Publisher's knowledge every effort has been made to contact relevant copyright holders and to clear any relevant copyright issues.
Any omissions that come to their attention will be remedied in future editions.

All rights reserved. No part of this book may be reprinted or reproduced or utilised in any form or by any electronic, mechanical, or other means, now known or hereafter invented, including photocopying and recording, or in any information storage or retrieval system, without permission in writing from the publishers.

Notice:
Product or corporate names may be trademarks or registered trademarks, and are used only for identification and explanation without intent to infringe.

BRITISH LIBRARY CATALOGUING IN PUBLICATION DATA

The globalization of space: Foucault and heterotopia.
1. Foucault, Michel, 1926–1984. 2. Space – Social aspects. 3. Environmental psychology.
I. Palladino, Mariangela, editor. II. Miller, John, editor.
304.2'3-dc23

ISBN-13: 978-1-8489-3462-7 (hbk)
ISBN-13: 978-0-367-59951-5 (pbk)

Typeset by Pickering & Chatto (Publishers) Limited

CONTENTS

Acknowledgement   vii
List of Contributors   ix
List of Figures   xi

Introduction – *Mariangela Palladino and John Miller*   1
Part I: State and Hegemony
   1 From Hegemony to Heterotopias: Geography as Epistemology
      in Gramsci and Foucault – *Mauro Pala*   13
   2 'An Occult Geometry of Capital': Heterotopia, History and Hyper-
      modernism in Iain Sinclair's Cultural Geography – *Tom Bristow*   29
Part II: Movement, Marginality, Containment
   3 Heterotopia and Placelessness in Brian Chikwava's *Harare North*
      – *Zoë Wicomb*   49
   4 'It's a Freedom Thing': Heterotopias and Gypsy Travellers'
      Spatiality – *Mariangela Palladino*   65
   5 Heterotopias of Illness – *Stella Bolaki*   81
Part III: Seas and Ships
   6 Writing the Littoral – *Abdulrazak Gurnah*   95
   7 Heterotopia and the Critical Cut – *Iain Chambers*   111
   8 'L'asile Flottant': Modernist Reflections by the *Armée du Salut*
      and le Corbusier on the Refuge/Refuse of Modernity
      – *Diane Morgan*   127
Part IV: Animals, Energy and Ecology
   9 Zooheterotopias – *John Miller*   149
   10 Soft Machines – *Fabienne Collignon*   165

Notes   183
Index   215

ACKNOWLEDGEMENT

Sincere gratitude is due to Roger Palmer, whose work was vital to the project's development.

LIST OF CONTRIBUTORS

Stella Bolaki is Lecturer in American Literature at the University of Kent. She works primarily in multi-ethnic American writing, Disability Studies and Medical Humanities. She is the author of S. Bolaki, *Unsettling the Bildungsroman: Reading Contemporary Ethnic American Women's Fiction* (Amsterdam: Rodopi, 2011) and is currently completing a new monograph titled *Illness as Many Narratives*.

Tom Bristow completed BA (Hons), MA (Hons) at the University of Leicester, and a PhD at the University of Edinburgh. Tom is a post-doctoral research fellow at the Australian Research Council Centre of Excellence for the History of Emotions, University of Melbourne; he is interested in British literature, ecocriticism, geocriticism, and the ecological humanities.

Iain Chambers teaches at the University of Naples, 'L'Orientale', and is known for his interdisciplinary and intercultural work on music, popular and metropolitan cultures. His books include: I. Chambers, *Migrancy, Culture, Identity* (London: Routledge, 1994); I. Chambers, *Culture After Humanism: History, Culture, Subjectivity* (London: Routledge, 2001); and I. Chambers, *Mediterranean Crossings: The Politics of an Interrupted Modernity* (Durham, NC: Duke University Press, 2008).

Fabienne Collignon is interested in American techno-culture and machine aesthetics. She is the author of F. Collignon, *Rocket States: Atomic Weaponry and the Cultural Imagination* (New York: Bloomsbury, 2014) and has published articles in *Textual Practice*, *CTheory*, *European Journal of American Studies* and *Configurations*.

Abdulrazak Gurnah is a writer, critic and scholar; he is the author of eight novels, the most recent is A. Gurnah, *The Last Gift* (New York: Bloomsbury, 2011). He is also Professor at the University of Kent (School of English). Among his publications, are: A. Gurnah (ed.), *Essays on African Writing* 2 vols (Oxford: Heinemann, 1995); and A. Gurnah (ed.), *The Cambridge Companion to Salman Rushdie* (Cambridge: Cambridge University Press, 2007).

John Miller is Lecturer in Nineteenth-Century Literature at the University of Sheffield. His first monograph, J. Miller, *Empire and the Animal Body: Violence, Identity and Ecology in Victorian Adventure Fiction* (London: Anthem Press, 2012), concerns depictions of exotic animals in Victorian adventure fiction. He also co-authored J. Miller and L. Miller, *Walrus* (London: Reaktion Books, 2014), as part of Reaktion's Animal series. John has taught at the universities of Glasgow, Edinburgh and East Anglia.

Diane Morgan is Lecturer in the School of Fine Art, History of Art & Cultural Studies, University of Leeds. Her publications include: D. Morgan, *Kant Trouble: The Obscurities of the Enlightened* (New York: Routledge, 2000); D. Morgan and K. Ansell Pearson (eds), *Nihilism Now!: Monsters of Energy* (Basingstoke: Palgrave Macmillan, 2000); and D. Morgan and G. Banham (eds), *Cosmopolitics and the Emergence of a Future* (Basingstoke: Palgrave Macmillan, 2007).

Mauro Pala is a scholar in critical theory and cultural and postcolonial studies, and has held research posts at Witwatersrand University, Johannesburg and Notre Dame University, Indiana. Since 2000 he has been Senior Lecturer in Comparative Literature and Literary Theory at the University of Cagliari.

Mariangela Palladino is Lecturer in the English Department at the University of Keele. Her research interests lie at the intersection of postcolonial literatures and cultures, diaspora, migration studies, and narratology. Mariangela has worked at the universities of Strathclyde, Glasgow, Edinburgh and Coimbra. She has published in *Modern Fiction Studies*, *The Journal of South African and American Studies*, *Interventions: Journal of Postcolonial Studies* and the *Journal of Postcolonial Writing*.

Zoë Wicomb is Emeritus Professor at the University of Strathclyde. She is the author of three novels: Z. Wicomb, *David's Story* (New York: Feminist Press at the City University of New York, 2001), Z. Wicomb, *Playing in the Light: A Novel* (New York: New Press, 2006), and Z. Wicomb, *October: A Novel* (New York: New Press, 2014); two collections of short stories: Z. Wicomb, *You Can't Get Lost in Cape Town* (New York: Pantheon Books, 1987) and Z. Wicomb, *The One that Got Away* (New York: New Press, 2008); and numerous critical essays on South African writing and culture.

LIST OF FIGURES

Figure 2.1: The Queen Elizabeth Olympic Park — 35
Figure 2.2: Road Closed Here, White Post Lane — 46
Figure 8.1: The 'without slum-dwellers' [*les sans-taudis*] make their way towards 'the floating asylum, the Louise Catherine' on the Seine — 128
Figure 8.2: The 'Cité du refuge' and its various operations — 129
Figure 8.3: Desperate, the social outcast is tempted to put at end to it all by drowning himself in the river — 131
Figure 8.4: 'The banquet of 800 seatings offered to those "without a slum" [*les sans-taudis*] of Paris' — 132
Figure 8.5: The Louise-Catherine at night — 133
Figure 8.6: Those who were invited to the Christmas meal — 135
Figure 8.7: 'A city where the poor will find his way' — 137
Figure 8.8: 'During the week of renunciation, please no abstention. The Factory of Good must continue its work' — 138
Figure 8.9: The radiating City of Refuge complex — 140
Figure 8.10: 'Our pupil officers contribute to the victory. Step after step the troops attack the mountain peak' — 142

INTRODUCTION

Mariangela Palladino and John Miller

Engagements with Michel Foucault's conception of heterotopia have routinely involved something approaching a disclaimer. For Edward Soja, Foucault's accounts of heterotopia are 'frustratingly incomplete, inconsistent, incoherent'.[1] To Peter Johnson, they are 'sketchy, open-ended and ambiguous',[2] while Hilde Heynen insists on the 'undecidability of the notion' of heterotopia.[3] It would be easy to continue in this vein, unfolding a litany of uncertainty and apparent hesitation that has dutifully accompanied the steadily expanding body of academic work on heterotopias since Foucault's first usage of the term in the 1960s. Ultimately, as Robert J. Topinka concludes, Foucault's delineations of the topic 'do not reduce to one succinct, unproblematic definition of the term, making scholarly attention to the topic difficult'.[4] Yet, and this is another almost obligatory observation on the Foucauldian heterotopia, the idea has proved endlessly appealing, even seductive. For a theme so incontestably ambiguous, the heterotopia has stimulated what might seem a disproportionate volume of critical responses, most prolifically from the social sciences. Johnson, in what is the most thorough summary of the topic to date, gives thirty-six examples of the 'dazzling variety of spaces that have been explored as illustrations of heterotopia' between 1990 and 2013, including 'underground band rooms in Hong Kong'; 'pornographic sites on the Internet'; 'the public nude beach' and an 'off-shore pirate radio station'.[5] It is difficult at a glance to formalize what these spaces have in common, except a somewhat general sense of being set-apart from what might contrastingly be thought of as the 'normal' or the 'everyday'; a sense that a heterotopia is a 'different' kind of space.

It is perhaps precisely the term's open-endedness that constitutes its allure. What Foucault's notion lacks in scholarly rigour it makes up for in suggestiveness. As James D. Faubian contends, 'Foucault's thoughts on heterotopia ... set aside the historian's usual empirical reserve'; there is an 'expansiveness'[6] to his writing on heterotopia and also, though this is not unusual for Foucault, a touch of poetry. As he somewhat mysteriously ends his 'Of Other Spaces', the essay that provides the conceptual centre of this study, 'In civilizations without boats, dreams dry up, espionage takes the place of adventure, and the police take the place of pirates'.[7] What is notable about this conclusion (and these are lines that will appear again in the chapters collected here) is not its thesis (even if in the conventional sense it might be said to have one), but its imagery. Foucault is operating here in between scholarship and creative writing; the essay is in itself an *other* kind of discursive space. The list of locations that Foucault himself cites as heterotopias (brothels, colonies, cemeteries, boarding schools, zoos, rest homes, prisons and more) is, as Johnson explains, 'almost mischievous in its variety'.[8] For all the serious-minded scholarly attention to heterotopias, the term begins playfully. Soja's summation of his essay's tone is perhaps the most appropriate: 'Foucault romps through the principles of heterotopology with unsystematic autobiographical enjoyment and disorderly irresponsibility'.[9]

The ad hoc way in which Foucault's theory of heterotopia has arrived in critical discourse may also contribute to a sense of intellectual glamour that has come to surround it. Michiel Dehaene and Lieven De Cauter explain that heterotopia entered 'architectural and urban theory … more as a rumour than as a codified concept'.[10] It is worth noting that 'heterotopia' is not in fact a coinage of Foucault's, but an adaptation of a relatively obscure medical term, denoting a body part 'occurring in an abnormal place'.[11] The Foucauldian heterotopia has in a sense two beginnings, though no real conclusion: as Arun Saldanha has it, the concept was more or less 'introduced and immediately abandoned' by Foucault.[12] His first published usage of 'heterotopia' appears in *Let Mots et Les Choses* (1966), translated in English as *The Order of Things* in 1970. The preface to this 'archaeology of the human sciences' departs from Jorge Luis Borges' invention in his *Book of Imaginary Beings* (1957) of 'a certain Chinese encyclopedia' structured around a surreal taxonomy of animal life in which creatures are understood through the following categories:

> (a) belonging to the Emperor (b) embalmed (c) tame (d) sucking pigs (e) sirens (f) fabulous (g) stray dogs (h) included in the present classification (i) frenzied (j) innumerable (k) drawn with a very fine camelhair brush (l) et cetera (m) having just broken the water pitcher (n) that from a long way off look like flies.[13]

Laughing uneasily, as he describes it, about Borges's fabular order of beings, Foucault offers 'heterotopia' as a mode of characterizing this disturbing conceptual arrangement. As he will do again, Foucault here juxtaposes heterotopia with utopia. Utopias 'afford consolation: although they have no real locality there is nevertheless a fantastic, untroubled region in which they are able to unfold'. Heterotopias, on the other hand, 'secretly undermine language'; they 'desiccate speech, stop words in their tracks, contest the very possibility of grammar at its source; they dissolve our myths and sterilize the lyricism of our sentences'.[14] This difficult passage reveals heterotopias as a disruptive force, one that operates here primarily linguistically (though importantly it is a language infused with spatial referents) by undermining an appearance or signification of a perfectly untroubled order. A heterotopia, then, is a space or a language in which to think things differently.

The second beginning of heterotopia can be located in a short talk for radio that Foucault delivered in 1966 on the subject of utopia. At the request of an architect, he delivered a lecture the following year based around the same material, though a published version would not arrive until 1984, with the English translation 'Of Other Spaces' appearing in *Diacritics* in 1986.[15] The essay opens by arguing for a historical shift between the nineteenth century and 'the present epoch'. If the 'great obsession' of the nineteenth century was history 'with its themes of development and suspension, of crisis and cycle, themes of the ever-accumulating past', then the twentieth century is the 'epoch of space ... the epoch of the near and far, of the side-by-side, of the dispersed'.[16] Space, of course, as Foucault goes on to elaborate, also 'has a history in Western experience',[17] but by the second half of the twentieth century it was already clear that 'the anxiety of our era has to do fundamentally with space' in a way that it had not been in previous periods.[18] By the time Foucault was writing, space, he suggests, had become a preeminent global concern.

Looking back from the vantage of the twenty-first century, Foucault's insistence on the intimate involvement of space and anxiety appears prescient. With the Earth's population passing 7 billion in 2011, questions of limits, territory and migration are more pressing than ever. The accelerating demand for dwindling resources in an increasingly destabilized global climate (both environmental and economic) offers a compelling reminder of the ultimate restrictions of our existence on a finite planet. As Bruno Latour pessimistically puts it, 'there is not the slightest chance for nature-and-society to be able to handle the crowding of organisms clamoring for a place to deploy and sustain their life forms'.[19] Spatial tensions increase social tensions; altered forms of hegemony transform strategies of resistance as new geographies of power are configured. The hyper-mobility of a globalized age finds a striking contrast in the restricted movements of those displaced by drought, famine and persecution. Indeed, the cosmopolitan, hyper-connected *citoyen du monde* and the refugee are perhaps the paradigmatic figures of our time.

The relationship between space and power, that other perennial concern of Foucault's, is perhaps the most urgent question of our time, and one that pertains to both intimate and macrocosmic scales. As Foucault wrote in a 1977 essay, 'The Eye of Power':

> A whole history remains to be written of *spaces* – which would at the same time be the history of *powers* (both of these terms in the plural) – from the great strategies of geopolitics to the little tactics of the habitat.[20]

Although undeveloped in its own terms, the conception of heterotopia does, then, clearly intersect with Foucault's wider project. As Nigel Thrift summarizes, 'Everywhere in Foucault there are ... markers of his sensitivity to spatial order – as a key to the constitution of power, as a marker of the self, as a requisite co-ingredient of numerous practices'.[21] Thinking about 'other spaces' (in their many forms), therefore, provides a conceptual tool for thinking about the geographical dimensions of the political and philosophical complexities of the late twentieth and twenty-first centuries.

Foucault's essay 'Of Other Spaces' focuses on a particular sense of the relationships between spaces; as he puts it 'the curious property of being in relation ... in such a way as to suspect, neutralize, or invert the set of relations that they happen to designate, mirror, or reflect'.[22] Heterotopology (the discourse, that is, of the heterotopia) is involved, then, in a simultaneous reflection and inversion of the mainstream world (for want of a better term) from which the heterotopia is removed. Foucault, again, insists on a connection between the utopia and the heterotopia. But while utopias are 'fundamentally unreal spaces' that present society in 'a perfected form', which is to say they reveal a 'direct or inverted analogy with the real space of Society', heterotopias are 'real spaces' and 'counter-sites' that 'remain absolutely different from all the sites that they reflect and speak about'.[23] Heterotopias should not, however, be understood in *opposition* to the space from which they are differentiated. They remain intimately involved with the rest of the world, even as they suspend its regulations and affects. Foucault extends this initial heterotopology through six principles. A detailed elucidation of these is beyond the scope of this introduction (they will at any rate return, variously, in the chapters that follow), but a brief summary of the principles is nonetheless useful.

Firstly, heterotopias are universal: 'there is not a single culture in the world that fails to constitute heterotopias'.[24] If this strikes a somewhat generalizing note, the second principle attempts to redress this: 'a society, as its history unfolds, can make an existing heterotopia function in a very different fashion'.[25] Heterotopias, while widely constituted, are culturally specific. Thirdly, more complicatedly, the 'heterotopia is capable of juxtaposing in a single real space several spaces, several sites that are in themselves incompatible'.[26] Giving the example of a Persian garden, said to 'bring together inside its rectangle four parts representing the four parts of the world', the heterotopia functions as a 'sort of microcosm'. The garden is both 'the smallest parcel of the world ... and ... the totality of the world'.[27] Foucault's fourth principle links space to time ; heterotopias merge with heterochronies ('slices of time' is Foucault's gloss on this awkward word) as they are understood to be constituted by an 'absolute break with their traditional time'. This might function in one of two ways. Heterotopias can be either, as in the examples of the museum or library, spaces in which time might to be said to accumulate in an archive which encloses all other times; or, as in the case of 'time in the mode of the festival', spaces in which time appears in its 'most fleeting, transitory, precarious aspect', as in such temporary emplacements as fairgrounds.[28] The fifth principle concerns 'a system of opening and closing' through which heterotopias can be sheltered or hidden so that they are only accessible under certain distinctive conditions.[29] Lastly, heterotopias have a 'function in relation to all the space that remains'. The heterotopia is set apart *and* connected; it exposes, reveals or recreates the 'real' spaces of society.

Sketching the principles of heterotopology in this brief abstracted mode leaves us with more questions than answers. It is in the specific *application* of Foucault's illusive idea that heterotopia becomes meaningful. The chapters collected here are responses to Foucault (some direct, some more tangential) that do not attempt the impossible (and not even necessarily desirable) task of finally rationalizing heterotopology into a coherent, explicit *system* of thought, but rather use the idea of heterotopia to explore specific spatial configurations, tracing flows of power, patterns of resistance, suspensions of normative order and, indeed, the re-emergence of normative order. The chapters take heterotopia away (slightly) from its better established territory in the social sciences into literary studies and critical and cultural theory. As a whole, the collection does not aim towards consensus or a standardized understanding of Foucault's essay, cultivating instead heterogeneous perspectives, styles and philosophical orientations that are unified by an engagement with heterotopias in the context of that most unwieldy and hotly contested of historical processes, globalization.

As a phenomenon often understood as the imposition or the emergence of a homogenous experience across the world through the expansion of a neoliberal economic vision, globalization invites, even demands thinking about different spaces. To reiterate, this is not to say that heterotopias should always be thought of as spaces of confrontation with or opposition against a dominant regime. Globalization also constitutes heterotopias; the tourist resort, for example, the most globalized of spaces, might be read in terms of heterotopology. Globalization produces new social (and material) spaces and, as Nancy Fraser observes, generates 'a new landscape of social regulation, more privatized and dispersed'.[30] As Jason Read writes in related terms in an essay on the 'genealogy of Homo-Economicus' under neoliberalism, economic production has moved from 'the closed space of the factory to become distributed across all of social space encompassing all spheres of cultural and social existence'.[31] Fraser's vision of the 'dispersed' landscape of social regulation and Read's identification of the 'distributed' character of production return us to Foucault's reflection on the 'epoch of space' as the epoch of the 'dispersed'.[32] Dispersal conjures up at once ideas of decentralization and of accessibility, an unbounded permeation of capital across global space. For all the ubiquity globalization implies, however, for Latour, 'When we speak of the global, of globalization, we always tend to exaggerate the extent to which we access this global sphere'.[33] The utopic space that globalization has supposedly extended for everybody to reach, is both limitlessly dispersed and also, in a sense, withheld from apprehension as a total regime.

Although we might conclude of globalization, as Latour does of ecological crisis (globalization's twin sibling) that in our current epoch 'the very notion of an outside is in doubt',[34] there are, of course, countless interruptions and evasions within the apparently global reach of capital. In this context, heterotopology becomes a valuable tool for both disclosing the homogenizing spatial effects of capital and making sense of 'other' spaces outside a hegemonic topography and a topography of hegemony. Johnson's conclusions about heterotopology, drawing on Maurice Blanchot, are here instructive. Heterotopology:

> is an attempt to think differently about, and uncouple the grip of, power relations: to overcome the dilemma of every form of resistance becoming entangled with or sustaining power. Heterotopias in this way light up an imaginary spatial field, a set of relations that are not separate from dominant structures and ideology, but go against the grain and offer lines of flight.[35]

Thus, heterotopias 'stage' – to borrow from Dehaene and De Cauter – the contradictions that society produces, they expose unsolved incongruities and interrogate established, real spaces. Heterotopias, to Dehaene and De Cauter, are placed at the 'intersection' of 'two axes': 'real/imaginary (utopia–heterotopia) and normal/other (topos–heterotopos)'. These make 'heterotopias into mirroring spaces'; the 'attachment to utopia charges heterotopia with the full ambiguity, even undecidability, of whether to attribute to it "eutopic" or "dystopic" qualities'.[36] Dehaene and De Cauter place heterotopias in an ambiguous, dialectic relation between real and imagined realms. This collection explores precisely this ambiguity by emplacing heterotopias in real spaces, yet as manifestations of imagined realities; as sites which resist reality and offer *other* ways of conceiving material/immaterial space.

The chapters of this collection, while engaging with a globalized understanding of space, forcefully and purposefully dis-engage from it: heterotopia is here presented as an indissoluble part of homogeneous, global spatiality, but also as a deviating energy which spins away from it. The contributions included here share an ethical commitment towards exploring and valuing *other spaces* as productive forces in generating novel conceptualizations of im/material space; these are not mere academic exercises, but have a sustained political strength. To Johnson's list of heterotopic spaces and applications of heterotopology, this collection adds: Gramscian hegemony, the Olympic Games, Gypsy Traveller sites, refugees in London, hospitals, Africa's East Coast, the Mediterranean, a floating asylum, the habitats of mythic creatures and windfarms. The chapters are organized into sections that reflect groupings of particular forms or arenas of the heterotopic (though connections also inevitably flow across the section boundaries).

Part I, 'State and Hegemony' investigates aspects of the experience of space in relation to economic and governmental forces. Heterotopias are considered here in relation to hegemonic sites – emplacements of power which do not necessarily materialize in one place, but are rather pervasive spatialities. The section opens with Mauro Pala's chapter, 'From Hegemony to Heterotopias: Geography as Epistemology in Gramsci and Foucault'. In this chapter, Pala develops a comparison between Foucault's and Gramsci's visions of space. 'It is undeniable',

Pala explains, 'that they often dealt with the same issues: from the genealogy of modernity and its characteristic institutions to the idea of nation, the link between power and knowledge ... [and the] radical distrust of totalising concepts'. The chapter explores how – in both Foucault's *oeuvre* and in Gramsci's philosophy of praxis – geography represents the methodological system to scrutinize hegemony. Pala examines Foucault's vision of space as a microphysics of power where the very notion of discipline is based on a meticulous parcellation of spaces. Gramsci's political theory too is marked by the use of the geographical, as he 'makes use of several of the famous metaphors which are part of a geographical conception of politics'. Starting from Foucault's geography and the inherent dynamics of repression and control at stake in any architectural product, Pala goes on to explore hegemony in Gramsci's work. Gramsci adopts a geographical perspective to analyze the Southern Question (a century old issue related to Italy's regional social, political, cultural and geographical heterogeneity). The Gramscian analysis of Fordism is also driven by a spatial approach whereby, as Pala observes, the factory becomes 'the production site that moulds the anthropological traits of an entire life system, and qualifies itself as a heterotopia'. In Gramsci's works, Pala contends, both examples of the Southern Question and the Ford factory as spatial strategies 're-evaluate the ability of each individual to influence or modify particular historical-territorial structures, while also identifying its oppressive or authoritarian nature'.

The relationship between heterotopology and hegemony is continued in Tom Bristow's '"An Occult Geometry of Capital": Heterotopia, History and Hypermodernism in Iain Sinclair's Cultural Geography'. Bristow is concerned initially with urban developments in the lead-up to the 2012 London Olympic Games. The Olympics, for Bristow, form part of the 'global capitalist culture industry'; their impact on the landscape of the Lower Lea Valley emerging as an act of enclosure most strikingly represented by the bright blue security fence that contained the works. If this site is a heterotopia, it is a heterotopia of social control that aims to subordinate historical and geographical specifics to a homogenous and hegemonic vision of the city produced in the service of what Bristow terms the 'heritage industry'. Contrastingly, the psychogeographic writings of the poet, novelist and critic, Iain Sinclair seek to destabilize authoritative voices, undermining the 'map of power' produced by global capital, instead seeking out 'unconscious drives ... and archaic energies'. In Sinclair's version we might understand heterotopia as a 'poetic site of resistance', or as a 'politicised space'. Central to Bristow's exploration of the economies of power at work in construction and reconstruction of London (both materially and ideologically) is a tension between Sinclair's work and that of the more conservative novelist Peter Ackroyd in whose account the Thames is read as a 'mirror for national identity'. Ackroyd's river is a 'unifying metaphor', while Sinclair's is a place of multiple, local histories. It is through these that Sinclair is able to critique the 'occult geometry of capital', London's managed flows (or 'pedestrian permeability'), 'monitored by a discreet surveillance technology' and underwritten by a supermarket chain so that Hackney in effect is transformed, as Sinclair explains, into 'a suburb of Tesco'.

Part II on 'Movement, Marginality, Containment' explores the interplay of movement and containment in the experience of marginalized groups and the places they inhabit. Involving the common denominator of resistance, the 'otherness' of sites such as cities, camps and clinics is both the product of normative (hegemonic) spatialization and a site of struggle. In 'Heterotopia and Placelessness in Brian Chikwava's *Harare North*', Zoë Wicomb explores some of the ways in which Foucault's concept of heterotopia intersects with postcoloniality. Arguing, following Johnson, that heterotopias do not necessarily 'subvert or support social systems', Wicomb reads the Zimbabwean author Brian Chikwava's first novel *Harare North* (2009) as an exemplification of 'a globalized age of migration' that engages with a Foucauldian spatiality without providing any 'hope of social change'. Wicomb is particularly interested here in how Chikwava's account of asylum seekers in London represents space in relation to subjectivity and how London (the Harare North of the title) appears as a 'place of illusion'. Chikwava's unnamed and unreliable first person narrator is an ambivalent figure whose identity appears to merge with that of his friend Shingi, part of a wider interest in doubleness and unstable selves that pervades the text. At the same time the novel's intertextual relationship with Sam Selvon's *Lonely Londoners* (1956) links the representation of Zimbabwean refugees with the earlier experience of West Indian migrants. While this connection 'speaks of the permanent lot of the migrant', the novel also asserts the difference of London between these distinct historical periods: 'London is and is not the same place'. In accordance, then, with Foucault's heterotopology, this is a city that finds its significance through relationships, both through time and across space . The interplay between sameness and difference that constitutes the narrator's experience renders this a 'fundamentally disturbing place' as 'the incompatible sites of Harare, contemporary London and London of the fifties are juxtaposed in a single real space'.

From refugees' relations to space, we move on to those of Gypsy Travellers with Mariangela Palladino's chapter. '"It's a Freedom thing": Heterotopias and Gypsy Travellers' Spatiality' explores how the mobility of these communities still remains profoundly misunderstood, criminalized, severely curbed and meticulously controlled. The narratives examined in this chapter – interviews with Gypsy Travellers and Mikey Walsh's 2009 memoir *Gypsy Boy* – offer paradigmatic examples of Foucauldian heterotopias, of 'other' spaces which deviate from normative spaces, an analysis with particular significance with regard to Gypsy Travellers' dwellings. Travellers' mobile homes recall several aspects of Foucault's heterotopian principles. A means of transportation as well as a home, a trailer, chalet or wagon defies dominant discursive categories, embodying the Foucauldian 'juxtaposition of incompatible terms'. Palladino investigates the contested nature of nomadic sites as a challenge to sedentary conceptions of movement, travel and space. Consequently, Gypsy Travellers' spatiality represents both a defensive mechanism against homogenizing, global processes and a novel way of conceiving of and inhabiting space. Resisting essentialized, hegemonic and sedentary spatialities, nomadism articulates *other* spaces of possibility.

Continuing Palladino's exploration of marginal sites, Stella Bolaki's 'Heterotopias of Illness' deploys heterotopology with regard to the spatialization of illness. This chapter focuses on real, tangible and habitable spaces, as well as on imagined spaces of escape, transformation or revelation. The 'real' spaces are the tuberculosis sanatorium, the cancer clinic and the dementia ward; for Bolaki, these medical spaces 'echoing scholars of heterotopias, could be seen as either "vulnerable and marginalized spaces" or spaces of "Other voices"'. Indeed, such sites have heterotopian qualities for they interrupt the continuity and normality of everyday ordinary space. The 'imagined' spaces are metaphorical heterotopic spatialities such as illness itself, as a condition, rather than a place. '[H]eterotopias draw us out of ourselves in peculiar ways'[37] and, as Bolaki demonstrates, illness 'does something similar, drawing us out of our familiar bodies and spaces, temporarily or sometimes permanently'. The chapter examines three illness narratives – *Madonna Swan: A Lakota Woman's Story* (1991), Audre Lorde's 'A Burst of Light: Living with Cancer' (1988), and Linda Grant's *Remind Me Who I Am Again* (1998) – and analyses space in relation to power, race, ethnicity and culture. Through these illness narratives, 'which conjugate both the material and metaphorical aspects of Foucauldian spaces', Bolaki investigates otherness and heterogeneity in medical spaces. Resonating with other chapters in the collection, this chapter also offers an analysis of heterotopias as sites of oppression and of resistance. For instance, Lorde's *The Cancer Journals* provides a powerful example of patient resistance in the refusal of a 'pink prosthesis' after her mastectomy; as a 'black lesbian feminist' her choice is a challenge to multiple forms of coercion and a resistance to the accepted image of a normative healthy body.

Part III, Seas and Ships, takes us onto water and onto land at the edge of water to investigate fluid geographies, flows of power across seas and between nations and the function of the ship, in Foucault's terms as 'the heterotopia par excellence'. Abdulrazak Gurnah's chapter 'Writing the Littoral' looks at the coast of East Africa through the work of Joseph Conrad, Karen Blixen, V. S. Naipaul and his younger brother Shiva Naipaul, picking up on the colonial and postcolonial resonances of heterotopia developed in Wicomb's chapter. Framing his chapter with Foucault's identification of the heterotopia as a site 'capable of juxtaposing in a single real space several spaces ... that are in themselves incompatible', Gurnah explores how the East African littoral becomes disembodied in colonial writing and overlaid with other meanings. To begin, Conrad's story of colonial traders in Mauritius, 'A Smile of Fortune' (1911) reveals a familiar 'strategy of self-consolidation through representation' that denies its location's complexity. An 'other space' (in this case the imaginative other space of colonial discourse) comes to be 'superimposed onto a real space', a pattern also evinced in Blixen's distortion of coastal cultures in *Out of Africa* (1937). The colonial literary conventions these texts disclose are also, Gurnah argues, 'rather alarmingly' continued in the writing of the Naipaul brothers. V. S. Naipaul's novel *A Bend in the River* (1979) is based to a great extent on a 'profound misunderstanding of the complexly hybrid community of the coast' that makes it 'reminiscent of colonial discourses'. Shiva Naipaul's travel book *North of South* (1978) contains a similar disengagement with Africa, taking recourse to 'the familiar clichés of imperial disdain', but does contain a 'moment of ambivalence' that shows the author at least 'grasped that there was more to what he was seeing than he fully understood'. Gurnah's chapter, then, witnesses an interplay and contest between the mythic and the real, and between homogenizing colonial voices and specific cultures, in these literary imaginings of the region.

Iain Chambers' 'Heterotopia and the Critical Cut' takes us into the Mediterranean, that 'marine cemetery of modern day migrant labour' in another chapter concerned with the (post)colonial dimensions of other spaces. While Gurnah focuses on the operation of imperial orthodoxy, Chambers is concerned with energies beyond a 'single-minded modernity' and with thinking that embraces the 'subjectivating forces, shifting combinations and unplanned vibrations' that do not 'remain locked in the power of established positions'. The sea with its indeterminate boundaries, its unavoidable, literal sense of flow, produces alternate forms of knowledge to those constituted by the homogenizing effects of global modernity, comprising 'an ontological challenge to the histories and events that apparently require terrestrial ground in order to be narrated'. Heterotopias 'spill ... out of the homogenous time of capital and nation' and invite us to 'drop deeper into the folds of the contemporary world'. Chambers is particularly interested here in music, 'an ecology of rhythms, beats and tonalities that produces sonic cartographies'. Music makes the hidden histories of the Mediterranean 'audible', an argument Chambers pursues with reference to the voice of Umm Kalthūm, an Egyptian singer of notable interest to Edward Said. Umm Kalthūm's voice expresses a 'distinct Mediterranean musicality' that also resonates with other musics: the blues, jazz, hip hop. Like the sea, music can propel us 'beyond the securities of territorial imperatives', inviting a critical practice 'sustained in the heterogeneity of times, rhythms and spaces, in the multiplications of modernity snapping the links of homogeneous understandings'. Global modernity, for Chambers, is always 'susceptible to unlicensed winds and currents' and 'alternative accountings of the world'.

With Diane Morgan's chapter 'L'Asile Flottant: Modernist Reflections by the *Armée du Salut* and Le Corbusier on the Refuge/Refuse of Modernity' we move into inland waterways. Morgan's chapter concerns the Louise Catherine, a coal barge converted by the great architect Le Corbusier between the wars and used as a floating asylum on the Seine by the French Salvation Army, the *Armée du Salut*. Made possible by a donation from Madeleine Zillhardt (the partner of the recently deceased painter Louise Catherine, after whom the boat was named), the vessel represents an unlikely alliance between 'Victorian Christianity and iconoclastic modernism' that disturbs stereotypical views of The Salvation Army. Not only did the barge provide a refuge for the homeless of Paris, it also, Morgan argues, constitutes a heterotopian space that destabilizes the 'habitual perceptions of the terrestrially bound'. The Louise Catherine served not just to alleviate the suffering of the urban poor before feeding them back into the labour market, but also offered a radical, even utopian message of social reform that extended beyond France to encompass the penal colonies of French Guiana. The work of the *Armée du salut* imagined a future 'order of things' that questioned prevalent, marginalizing ideas of 'the criminal' and the socially dispossessed. Le Corbusier's thought appears to be supportive of the *Armée du salut*'s idealistic message, but is also, Morgan explains, problematically entwined with the rise of fascism, an association that disrupts the sense of the boat as an unequivocally radical space. Morgan concludes by returning to Foucault and the distance between utopia and heterotopia, arguing that exploring the heterotopian character of the asylum with its 'heterogeneous, nonconformist and transient population' can instead be part of a history of utopian thinking that undermines a totalitarian requirement for 'homogeneity, normality and conformity'.

Part IV 'Animals, Energy and Ecology' brings an environmental perspective to bear on globalization and heterotopology. John Miller's chapter 'Zooheterotopias' focuses on the development throughout the twentieth century of the pseudoscience cryptozoology, the study of animals usually thought to be mythical or extinct. Such a belief that the world may still contain creatures not formally recognized by science is evidently based on an understanding of the resilience of space (or more accurately habitat) in the era of global capital. While the usual assumption is that the world's secrets are now exhausted by centuries of exploration and agricultural and industrial development, cryptozoology holds on to the possibility that mysterious domains may remain. Taking Arthur Conan Doyle's *The Lost World* (1912) and *King Kong* (1933, both the movie and Delos W Lovelace's novel) as emblematic texts of cryptozoology, Miller explores three of Foucault's principles of heterotopias as *zoo*heterotopias. Miller's first zooheterotopic principle is that discourses of space in this context are also temporal discourses. These are, in Edgar Rice Burroughs' famous title, lands that time forgot; as such these are spaces that appear resistant to the spread of global capital. Secondly, entry to the zooheterotopia is governed by a kind of ritual, connected to Foucault's sense that 'heterotopias always presuppose a system of opening or closing that both isolates them and makes them penetrable'. While this motif sets the zooheterotopia apart from the rest of the world, it also revolves around heroic performances of masculinity that reveal the involvement of cryptozoology with imperialism. Lastly, Miller explores the relation of the zooheterotopia to all other sites through the insistence that the lost creature must be commodified into global capital through the monster's return to the industrial metropolis. Ultimately, Miller concludes, the cryptozoological imagination both bemoans and conversely becomes part of capitalist modernity.

The final chapter, Fabienne Collignon's 'Soft Machines' investigates the wind farm as heterotopia. Like Miller, Collignon is interested in the ways in which discourses of the environment are involved with global capital, particularly here with capitalist technoculture. Collignon, examines 'the wind turbine and wind farm as both symbols of utopian thought and marked out or marketed "countersites" to entropy-inducing fossil fuel and nuclear technologies'. But while these machines might be understood as 'soft' for the sense of benevolence they are usually taken to communicate, they also 'resolutely reaffirm the dominance of conglomerates' who market their green credentials while remaining invested in the status quo. In a chapter that draws on Fredric Jameson, Marshall MacLuhan and Roland Barthes, in addition to Foucault, Collignon's analysis draws on the materiality of the turbines themselves' – how 'the turbine's thrust demands a techno-fetishist attention' – and how windfarms have been represented in popular culture, including, for example, Christopher Nolan's 2010 film *Inception*. Not only is the rhetoric of sustainability often inclined to be 'war-like' (as in Al Gore's" 'strategic environment initiative'), but the machines themselves evoke a technological sublime associated with fascism (there are echoes here of Morgan's chapter). Placing the turbine blades within a 'semiotics of plastic', Collignon traces their debt to Cold War militarism so that for all the green credentials of these heterotopias, in the end 'the turbine for the most part forms another device that legitimises the established regime'.

A final heterotopia that haunts these pages is that of the academy itself. This collection has been conceived across continents and time zones and realized in a virtual space of instantaneity (the non-place of Skype). Where once the academy was widely represented in that most heterotopian of images, the proverbial ivory tower, it is impossible now not to acknowledge the growing containment of the contemporary academy within a discourse of global capital that insists on an increasingly 'flexiblized' intellectual workforce. To make sense of this disquieting phenomenon, Marina Warner in a recent article in the *London Review of Books* borrows the terminology of the expressionist architecture witnessed in the buildings of her former employers, the University of Essex. For Warner, new brutalism 'in academia was taking on another meaning'.[38] The current model for higher education 'mimics supermarkets' competition on the high street'[39] and weakens heterogeneous realities in the face of a wider, global, international project. The hegemonic spatial formations shaping and pervading academia today are only a reminder of the Foucauldian space-power binomial; as Foucault put it, 'space is fundamental in any exercise of power'.[40]

1 FROM HEGEMONY TO HETEROTOPIAS: GEOGRAPHY AS EPISTEMOLOGY IN GRAMSCI AND FOUCAULT

Mauro Pala

Strategies and Spatial Narration

In 1976 Michel Foucault called upon the geographers of the *Herodote* journal to give an answer to whether the notion of strategy was essential for knowledge, and how strategy transformed relations of power into forms of domination:

> the notion of strategy is essential if one wants to make an analysis of knowledge (savoir) and its relations with power. I am asking myself, that I address to you, thinking that you are without doubt more advanced than me on this path. Does it necessarily imply that through the knowledge in question one wages war? Does strategy not allow the relations of power as techniques of domination? If I understand you correctly, you are aiming to constitute a knowledge of spaces (un savoir des espaces). Is it important for you to constitute it as a science?[1]

The very way in which these questions were posed, suggests the importance Foucault attributes to space in the articulation of power, omnipresent in his theories and likened to a constantly moving form of energy. In keeping with the exigency of the matter, these experts tried hard to question the sense of their discipline, coming up with an almost unanimous reply, which was explicitly political in its nature: different strategies produce different spaces, clearly implying different consequences, even at the level of new historical configurations. More precisely, Claude Raffestin explains that: 'in geography, the knowledge linked to the scientific analysis has been transformed into strategy and, very precisely, into a technique of dominating and even occupying economic political and geographical space'.[2]

The passage from knowledge to the techniques of dominating recalls the short but intense *Les Hétérotopies* where, starting from the modern condition, the philosopher had already proclaimed the political result of space and how it acquires territories: 'we are in an epoch in which space is given to us in the form of relations between emplacements'.³ Underlining the relational and functional nature of space, Foucault proposes other spaces as an alternative to reality, recalling the potential of utopias through the mirroring mechanism, and at the same time, investigating the structure of what exists, through the 'system of opening and closing' which is actuated by heterotopias.⁴ Focusing on 'emplacement' as a set of spatial and temporal coordinates, he shows an interest in space, not only as an alternative to history, but also as its complement and corrective. Already in his 1964 conference on *Le langage de l'espace*, Foucault's vision of space had nothing to do with any kind of 'spirit of time' generalizations, but was instead a microphysics of power, the very means that allows history to develop through 'implications and insights concerning spatiality'.⁵ Clear evidence of this is demonstrated in how Foucault builds 'his' history drawing upon spatial relations: *Madness and Civilization* (1967) sees the shaping of a 'geography of haunted places';⁶ in *The Birth of the Clinic* (1976) spatialization is applied to forms of diagnosis and the location of clinics, while in *Discipline and Punish,* the very notion of discipline is based on a meticulous parcellation of spaces. Finally, *Panopticum* is the perfect synopsis of a hermeneutic system and its presuppositions, as well as the introverted proof of the existence of research into this system, seeing that an actual place, an architectural product, with all its planning implications, is both the form and instrument of the dynamics of repression and control, in force in Europe from more or less the seventeenth century onwards. So where does this hypothesis on the functions of heterotopias spring from, why all this interest in the sense of strategies? Perhaps spatial coordinates are needed to substantiate the history that is taking shape in Foucault's work: a history that wants to be rebuilt from below, on a local basis which is intrinsically unstable, and whose structure appears to be methodologically founded on episodes which at first glance, seem to be of secondary importance, but then, thanks to their long *durée,* come to represent the trends of an era, trends that recall Braudel's great geographical panorama.

Notwithstanding an obvious spatial expansion, we are dealing with a history that is unable to grow *vertically,* both because of an inherent discontinuity and also due to the aporias that result from Louis Althusser's *coupure épistemologique.* The works of Marc Bloch and Lucien Febvre on Braudel, and those of Bachelard and Georges Canguilhem on Althusser, play their part in creating Foucault's conceptual constellation. They do not only overcome the idea of causality and continuity, but also what disappears is the axiom of a transcendental subject able to explain the sense of events, permitting history to freeze 'documents into monuments'.⁷ Putting aside the grand narratives, whose epigraph would be written by Lyotard, Foucault also rejects the synthesis of structuralism, and deliberately hinges his geography on a condition of possibility, within which the forms of knowledge are articulated time and time again:

From Hegemony to Heterotopias: Geography as Epistemology in Gramsci and Foucault 15

> What Foucault calls the archaeological level or system consists in a set of rules of formation which determine the conditions of possibility of all that can be said within the particular discourse at any time. Language ceases to be a universal medium of representation and becomes itself a historical phenomenon.[8]

Such conditions are summarized by the idea of an episteme which, despite its obvious relation with a particular historical or geographical field, does not intensify its nature, but merely limits itself to stating what can be said and how it should be said. An environmental cataloguing, whose mapping out, however, actually records a system which is the result of a complex strategy.

And indeed it was this insistence on strategy, and on the ways in which a territory can be socially negotiated that provided the inspiration for this comparison, albeit a limited one and not just for editorial limits, between Foucault's and Gramsci's conception of geography. Notwithstanding the historical and hermeneutic distance between these two key thinkers of the twentieth century, and the diverging results of their epistemological positions, it is undeniable that they often dealt with the same issues: from the genealogy of modernity and its characteristic institutions to the idea of nation, the link between power and knowledge, which both scholars perceived in the linguistic phenomena that characterize hegemony and discursive formations. They also share a radical distrust of totalizing concepts and an imminent sense of crisis as a feature of contemporaneity; Foucault wishes to highlight how certain groups have been environmentally and ethically marginalized, while Gramsci studies the ways in which a certain social class can achieve hegemony in a determined social context: in all these cases, the dynamics which underlie these phenomena are of a marked geographical nature.

However, no matter what ideological burdens distinguish them, and notwithstanding the interfacing of these themes, and an analysis based on their textual and discursive nature, as we shall see, their final interpretation is actually quite different. This is mainly because of their unprejudiced views on the possibility and possible role of a social agent, meaning that they do not follow any pre-established ethical system.

The essay on heterotopias is a good starting point for this comparison: in fact, this is where Foucault, skillfully making use of the typical rhetoric of fairy tales, alludes to the potential that places already have, and which they can acquire thanks to the interventions they undergo: this investment, which is also a parameter for the analysis and the 'will to knowledge' that drives power, operates both in individual places and at a national level, finally projecting itself on a global scale in the relation between different civilizations. Furthermore, the resulting modalities of exclusion actually make up a watershed between the Classical and pre-Classical periods, but in any case, whatever the scale of this mapping, it stimulates a 'gray, meticulous, and patiently documentary'[9] validation of what exists. Foucault creates an epistemology structured in two concomitant, but different ways; firstly, an archaeological path that highlights the power-knowledge relation, while the second meaning, considers space as a series of points where this selfsame power is applied, namely, in architecture, town planning, in clinics and prisons.[10] Thus, we are dealing with a multi-level spatial analysis, which is, however, always dominated by the space of the power-knowledge relation; a level is made up of an urban space, that is to say, the spaces of social life where forms of organization and control are most evident, including the sharing and distribution of power at the level of the human body. Human bodies are themselves an individual sphere, but they are interconnected with the environment, which touches on the Deleuzean approach to space. Indeed, Deleuze started out with his post-1968 *Anti Oedipus*, in a familiar environment corresponding to the subconscious, to then project himself into the collective dimension of *A Thousand Plateaus,* using etiology to face up to a series of territorial concatenations.[11] Finally, Foucault also has a type of space, where expectations, prohibitions and achrony occur, and space becomes the necessary condition for power, so that this emanation produces an extensive review of geographical notions:

> territory is no doubt a geographical notion, but it's first of all a juridical-political one: the area controlled by a certain kind of power. Field is an economic-juridical notion. Displacement: what displaces itself is an army, a squadron, a population. Domain is a juridical-political notion. Soil is a historic-geological notion. Region is a fiscal, administrative notion. Horizon is a pictorial, but also a strategic notion.[12]

All these different denominations converge into a heightened perception of a dynamic and political conception of geography.

In this famous interview, Foucault does not just voice his criticism of a practically static concept of geography, but introduces what Chris Philo defines as a 'theoretical input' into the discipline.[13] As a result of this, while the prevailing idea of history is dismantled in the background, geography itself becomes a source and an antidote for the immovability of historical or scientific discourse, thanks to its ability for highlighting local variations that reconsider and contradict the continuity and the coherence of these subjects. Foucault believes that historical context must go beyond the mere relativization of the phenomenological subject, which is abandoned once and for all, in favour of an analysis that corresponds to the concept of genealogy:

> And this is what I would call genealogy, that is, a form of history which can account for the constitution of knowledge, discourses, domains of objects, etc. without having to make reference to a subject which is either transcendental in relation to the field of events or runs in its empty sameness throughout the course of history.[14]

This is the inspiration for the Foucauldian project of a 'general' history that does away with the founding principles of conventional history, known as 'total' history: it is no longer, therefore, either the synoptic description of a civilization, or a corpus of laws that sanctions its interior coherence; it is not the radiant aspect of a historical period, but rather a 'general history which would deploy the space of dispersion'.[15] Traditionally, historians have always paid little attention to social geography, which for Foucault corresponds to a phenomenology of the events themselves, but without any superimposed order. On the contrary, the archaeology of knowledge actually operates on the element that historians have traditionally eliminated from their construct, namely discontinuity:

> Thus, in place of the continuous chronology of reason, which was invariably traced back to some inaccessible origin ... the notion of discontinuity assumes a major role in the historical disciplines ... discontinuity was the stigma of temporal dislocation that it was the historian's task to remove from history. It has now become one of the basic elements of historical analysis. First, it constitutes a deliberate operation on the part of the historian ... for he must, at least as a systematic hypothesis, distinguish between the possible levels of analysis, the method proper to each, and the periodization that best suits them.[16]

This is the way that Foucault not only erases any historical teleology, but also frames the historian's activity within a precise spatial context, where he is able to detect forms of power.

According to Foucault, knowledge is actual power, and not just a reflection of the power relations that exist within institutions: for this reason, some aspects of reality may or may not be presented as natural in his proposals, in a context where the basic idea is that power is generally concealed, despite its pervading presence. The Foucauldian link between power and knowledge puts forward a whole series of questions that, as is clearly shown, directly involve the production of geographical knowledge by, for example, feminist geography, postcolonial studies and critical geopolitics. His geographical approach recalls a hermeneutic system within which the role of interpreting the judgment is in turn subject to scrutiny, thereby eliminating the mimesis between whosoever judges, and what is being judged. This approach is based on the assumption that interpretation is a dialogue between the collected data from a specific field and context, and the researcher, who in turn necessarily represents a particular institutional and cultural environment.[17] In this study environmental and geographical conditions take on unprecedented importance for their distinctive irregularities, which refute an unequivocal interpretation of any sort of spatial temporal knot. Such an analysis causes mirroring in heterotopias to surface once again, reverberating on the works of the inventor and instigator of this project, encouraging him to gain a deeper understanding:

> I am over there, there where I am not, a sort of shadow that gives me my own visibility, that enables me to see myself there where I am absent: such is the utopia of the mirror. But it is also a heterotopia in so far as the mirror does exist in reality, where it exerts a sort of counteraction on the position that I occupy.[18]

Michel Serres observes that 'the mirror symmetries and complementarities have their own reason. This is that the thinking subject always finds himself on the other side of the dividing line'.[19] This is how Foucault violates the classical convention that sees the knower and the known as separate, postulating a gnoseological perspective in its place, where the knower is aware of the artificiality of the conceptualization in which he/she has played a part.

Foucault, just like Gramsci, makes frequent use of metaphors that are intended to define power relations as 'strategic clashes'[20] between opposing fronts, but the French philosopher is not so concerned with the battle itself, as much as with its preceding justifications and subsequent interpretations. The objective of Foucault's scrutiny does not coincide with the examination of the reasons behind the success or failure of a certain plan, but rather is focused on how to 'discover the point at which these practices became coherent reflective techniques with definite goals',[21] to then denounce their ideological connotation, as they are being disclosed. Equally, Foucault has never dwelt upon the founding principles of his spatial conceptualization, adopting a sense of abstract space which gives rise to a perception of dispersion that is always at a local level, inasmuch as it is a procedure that evades every kind of hierarchy or system. Nonetheless,

> imagining Foucault as less the geometer of power and more the patient archaeologist of substantive geographies is something that apparently resonates with his own views, notably when he highlights the value of proceeding with a clear attunement to real, worldly spaces ... all jumbled up together and related to another through spatial relations.[22]

Geography through the Philosophy of Praxis

Understanding Gramsci's relationship with space means starting from the link between the Sardinian intellectual and philosophy. However, compared to Michel Foucault's basically mechanistic usage of terms found in the opening pages of his essay on heterotopias, Gramsci's philosophy is fuelled by a history that is seen in rather more complex and dynamic terms. In fact, history and geography are part of what Frosini[23] defines as the material nature of the *Prison Notebooks,* and this nature can only be appreciated after a philological reading of the work where

> Philology is the methodological expression of the importance of particular facts seen as a defined and specific individuality. This method is in contrast with the 'large number theories' or 'statistics' of all or some of the natural science.[24]

Gramsci demonstrates how Marx exposes the ideological nature of idealism by highlighting the link between philosophy and historical reality simultaneously, and thereby revealing history's point-blank political nature. Therefore, philosophy is not a freestanding sphere, but when all is said and done, it is modified and constantly put to the test by historical events, in a reversal of the normal path of historicism, and the true essence of the *Quaderni* clearly emerges from this dialectic between history and philosophy. For Gramsci, the *Prison Notebooks* represented a complex stage of constant re-elaboration, which was not just an evolutionary process but also a moment to ponder on the past, and the *Quaderni* helped him to understand 'the concept of unity and practice of philosophy and politics'.[25] The autobiographical style of the *Quaderni* and the narration therein, have been likened by Alberto Burgio to a great history book:

> the history of the Western bourgeois world, or as Gramsci simply puts it, of the 'modern world', a critical history of European modernization: a huge book of history which is, at the same time – and here we find the first methodological similarity with Foucault – a book about history, its meaning and its logic, its relationship with philosophy (theoretical thought), and in the verification stage, about geography.[26]

This critical re-reading of modernity is brought about by philosophical praxis, although Gramsci's definition goes beyond that of Marxism and its links with idealism, as interpreted by several of its leading exponents, such as Croce, Sorel and Bergson. Furthermore, adopting a philosophical *praxis* wrong-foots the historiographical configuration of a teleological kind, present in many of Marx's readings, and restores the traditional 'scission between high and low culture, between intellectuals and the people, which gives rise to the scission between theory and practice and the consequence of the division of society into classes which has characterized the story of mankind'.[27] Within the field of historical materialism:

> philosophy is ... the full consciousness of contradictions, the consciousness wherein the philosopher himself, understood both as an individual and as a social group, not only understands contradictions but posits himself as an element of the contradiction and raises this element to a principle of politics and action. 'Man in general' is negated, and all immutably 'unitary' concepts are derided and destroyed, insofar as they express the concept of a 'man in general' or of 'human nature' immanent in every man. But even historical materialism is an expression of historical contradictions; indeed, it is perfect, complete expression of such contradictions ... The philosopher today cannot evade the present terrain of contradictions; he cannot affirm – other than generically – a world free of contradictions without immediately creating a utopia. This does not mean that utopia is devoid of philosophical value, for it has a political value, and every politics is implicitly a philosophy.[28]

Here, Gramsci is negating the idealistic concept of man and, at the same time, he is reaffirming the specific spatial conditions through which he is shaped. In this way, Gramsci is also denying the consolatory or unitary concepts of man and human nature through a continual inspection of historical conditions, giving a marked dialectic character to the perception of reality, fruit of Hegel's speculative thought:

> In other words, Gramsci is saying that one cannot think of reality, perceived in its contradictory character (which as we shall see extends over several levels) if not in a way of thinking that this contradiction elevates to the principle of knowledge.[29]

Unlike other Marxist thinkers, such as Bucharin and Lukacs, Gramsci does not see any possible synthesis or harmony in reality, but regards it as an endless crisis.

This politicization of philosophy and the related negation of theological historicism is followed by a lexical examination of political territory; at this stage, Gramsci makes use of several of the famous metaphors which are part of a geographical conception of politics. This idea ranges from the 'war of maneuver' to a 'war of position', to the antimonies of North and South, East and West, of the two ideologically opposed, but actually osmotic spheres of City and Country. Quoting Valentino Gerratana, the editor of the first critical edition of the *Quaderni*, Guido Liguori[30] reaffirms that Gramsci's method is more analogical than analytical, which obviously favours the use of metaphors especially in a geographical context; even though metaphors may partially distort the concept being expressed, their great versatility and inherent innovation are also capable of portraying both the ample nature of the analysis, as well as the 'liquid' substance of the topic in question. In Gramsci, in complete contrast to Ranke's historicism, institutions are continually exposed to the destructive dialectic of space animated by social agents. 'The idea of the state as an integral structure corresponds to an interpretation of modernity as an era characterized by a process of fluidification and integration of social bodies.'[31] Just as in Foucault, modernity is launched under the aegis of the bourgeoisie, although its dynamism is rather more stratified than that of the French scholar, and heralds articulated social cohesion, both in terms of the control of productive processes and of ideological direction.

But even before the *Quaderni*, Gramsci had already started to analyze human geography, and the 1926 *Alcuni temi della Questione meridionale* (Some Issues of the Southern Question) represents the first stage in this development. The work deals with the age-old and still unresolved controversy over the unification of Italy, which had started with the annexation of the former Bourbon Kingdom of the two Sicilies by the House of Savoy, which proclaimed itself Kingdom of Italy in 1861, without, however, adapting its constitution and relative institutions to the regional heterogeneity that characterized the Italian peninsula. Indeed, just a short while after unification, several liberally minded politicians with close links to the government, had denounced how the process of national construction in the south could be likened to a kind of forceful occupation, a sort of internal colonization. Just before his imprisonment, Gramsci wrote that the matter should be tackled 'as a territorial problem, that is to say, as one of the aspect of the national dispute'.[32] For Gramsci, any kind of intervention on a territorial basis meant dismantling the rhetoric of Risorgimento style national unity, while simultaneously denouncing the drain on resources which damaged the south and favoured the industrialized north. But the specific economic aspect, fruit of the alliance between the southern landowners and the northern industrialists, falls within a geopolitical analysis that goes beyond a simple bipolarism, to be organized on all levels, and whose starting point is the actual discontinuity and incongruence in the debate on Italian unification. Using words that are quite similar to Foucault's, Gramsci defines the South as a 'massive social disintegration' and hopes for 'molecular' intervention'[33] on what appears to be a geological stratification of three interconnected levels, from the 'amorphous' masses of the peasantry to the rural bourgeoisie, right up to the great intellectuals who like Croce, in their role as mediators, dominate this complex historical block. Gramsci's analysis then moves from the social structure of the south to the prejudices – Foucault's discursive formations – accompanying Northern rule, to the 'complex feeling that has arisen in the North towards the South'. This meant that the 'poverty' of the South had no historical explanation for the popular masses of the North, who could not conceive that this unity had not been created on an equal basis, but rather, 'as a hegemony of the North over the South in the city-country territorial relationship; in other words that the North was practically a "parasite" getting rich at the expense of the South'.[34] The specificity of this analysis at a territorial level confirms that passive revolution seen as a form of ideological rule, or rather, ongoing hegemony 'was not simply a metaphor but was constitutive of the actuality of spatial social relations'.[35] From the 70s onwards, the productive power of the 'Southern Question' seen as a form of 'Orientalism in One Country' – the title of the brilliant anthology of essays edited by Jane Schneider[36] – is quite staggering: Guha, Said, and recently Chatterjee and Spivak have acknowledged this essay with the right to primogeniture of a whole series of extensively covered topics in subaltern studies.

And once again Gramsci starts from the Southern Question and defines intellectuals as acting first as 'intermediari' [intermediaries] and then as a block in their own right. He then considers 'the limits of the term's significance',[37] and tries to understand exactly what the field of action of intellectuals is, or rather, whether this field is limited to a group or is instrumental to a whole class. This research not only results in 'a broadening of the concept of the intellectual',[38] but above all underlines the functionality of the intellectual with regards to different geographical circumstances. The Italian intellectual is often unable to come to terms with his own land due to the 'function the intellectuals had had in medieval cosmopolitan environments for the fact that the Papal See was in Italy'.[39]

As can be appreciated from this outline, Gramsci also starts from the differences, and just like Foucault, his analysis becomes more and more wide-reaching and increasingly significant, and it is here, in this specific case, that the reality of the South provides an explanation for the problems of the Risorgimento; various factors are interwoven within the Southern Question, seeing that internal and international capitalist relations mould both economy and state. 'In an Italy which is evermore integrated into a developed capitalist Europe ... the Southern Question is thus, to a certain extent, also the Italian way of being in a determined geographical area'.[40]

When, in the late 19th century, Gramsci realized that capitalism in all its manifestations had already generated a new social sphere, he also devised a broadening of the concept of State, underlining the increasingly close relationship at a political and economic level and in an intensification of the reciprocal link between these two spheres; 'he is proposing a radical interpretation of power in contemporary society'[41] within which 'civil society represents the political and cultural hegemony of a social group over the whole of society'.[42] To explain how the private sphere, represented by the interests of the ruling classes, permeates the State, he calls on Hegelian Associationism, through which Gramsci's concept of hegemony defines an albeit temporarily surrendered territory as homogenous and identifiable from the type of power exerted on it. In a steady progression, starting from the power of the ruling classes within a certain State, one can arrive at the nature of that state, and gradually, classify the history of states: 'the historical unity of the ruling classes is found in the state, and their history is essentially the history of states and groups of states. This unity must be concrete, hence it is the outcome of the relations between the state and civil society'.[43]

Apart from the state's occupation of civil society, the choice of a spatial prospective to scrutinize the entire Southern Question and its exemplary nature, makes it clear that another of Gramsci's famous contributions, namely the passive revolution, does not simply record the state of things in an impersonal way, but rather, focuses on an extremely variable relation between opposing factors, bringing about just as many hegemonic outcomes as settings – in the geopolitical sense – where a hegemonic expression can occur. In modernity, 'political and cultural hegemony'[44] replaces the pre-modern forms of direct rule, supporting and speeding up the 'expansive' bourgeois phase between 1789 and 1870. At the same time as this 'broadening' of the State which ends with the so-called permanent revolution, the 'massive' structure of modern democracy with its 'trenches' and 'permanent fortifications' is strengthened between 1789 and 1848, giving rise to that war of position that characterized the world scenario after the Russian revolution. On a global level, the situation was completely different in the colonies and economically backward countries, because they lacked a bourgeoisie 'in perpetual motion', capable of absorbing the whole of society in its cultural and economic sphere.[45]

Just as in Foucault, knowledge is the instrument of power even in the geographical perspectives involved in the Southern Question, but the circulation of knowledge comes about through the action of intellectuals, who can be identified within a system strategically directed at a clear goal; different groups of intellectuals in different periods build the hegemony under a form of intellectual and moral leadership, within which consent and persuasion make for the re-elaboration of a 'conception of the world' to be adopted as the dominating vision inside certain spatial temporal circumstances. Even here this vision of the world is not a concept of an idealist kind, but is linked to the space that is transformed into territory, the result, according to Edward Said, of a 'very powerful geographical sense'[46] or rather, made possible by a careful examination of the relations between 'nature and society'[47] in view of a possible transformation. Hegemony in this spatial variation is endowed with heterotopic potential meaning that it cannot exist in the absence of territories and relative geographical conditions where these social changes can occur.

As the twenties drew to a close, the internationalization of the Southern Question as an extension of this bourgeois hegemony is realized with Americanism which, compared to Europe and Italy, represented a horizon for possible development and as such, could have even become a sort of heterotopia for the boost to rationalization that set it apart. Gramsci is not so interested in America as a real place, as much as in this evolution, in the fact that a modality or lifestyle is taking shape, an Americanization that could spread throughout the West. The Ford factory confirms that 'you can't talk about national without territorial'[48] which makes it the summary and the epicentre of a vast review of an exemplary amount of meaningful experiences. When dealing with Fordism, or rather, the passage from the old economic order to a planned economy, Gramsci adopts a genealogical view, where 'genealogy is thus situated within the articulation of the body and history'[49] and this explains why, in the name of programmability, the legitimacy of the private space is sacrificed, as the factory 'desacralizes' the borders[50] between family intimacy and the public sphere, fusing, just like in the emplacement of the heterotopias, productivity with demography; hence, the 'preliminary condition'[51] for Americanization is the adjustment of the population to 'a programmed economy'.[52] 'Hegemony is born in the factory and does not need many political and ideological intermediaries'[53] or rather, it is the production site that moulds the anthropological traits of an entire life system, and qualifies itself as a heterotopia. It is not exempt from dystopian outcomes, since it can be assimilated to a 'kind of effectively enacted utopia in which the real sites, all the other real sites that can be found within the culture, are simultaneously represented, contested and inverted'.[54] This is the exact heterotopic feature that Gramsci clearly defines when he states that Ford's organization of the manufacturing plant represented the last stage[55] in an evolution of human society 'ever since nomadism'. This heterotopic mutation of the factory passes through a 'regulation of sexual instincts'[56] as the final stage of a more general 'coercion' to which Puritanism, encouraging the emulation of the wealthy classes, has contributed in defining a 'superior type' of monogamous worker, who 'does not squander his nervous energy in the messy and exciting search for occasional sexual satisfaction'.[57] The internalization of the 'family in the broad sense', that is to say, of 'a new form of sexual union' centred on monogamy and relative stability[58] is brought about through a 'form of coercion of a new kind' that 'can be none other than self-coercion, in other words, self-discipline'.[59] Despite the fact that the Ford factory was the example of a national situation, the spaces where the described phenomena took place would always be limited, and this is confirmed by the nature of 'disciplines as techniques developed in well circumscribed places ... which sought to increase the productivity of the individual while reducing his potential for insubordination – in other words, to place in an inversely proportional relationship the individual's economic profitability and his political autonomy'.[60] We are dealing with a process of masochist subjection (*assujettissement*) and objectification (*objectivation*) which intervenes directly on the bodies of those whose subjection is assured, imposing 'a relation of docility – usefulness' upon them.[61] On an analytical level, the analogies between the American factory and the intuitions in the *History of Sexuality*[62] are made clear, when Gramsci stresses the 'the progress in hygiene, which has raised the average life expectancy and continued to place the sexual question as a fundamental aspect, separate from the economic one'.[63] Both Foucault and Gramsci base the relation between knowledge and power, as clearly shown by the Ford factory, on local contexts and verifiable mechanisms in a concrete reality; places which, in their turn, provide a frame for and exemplify the social dynamics resulting from a certain historical contingency. Both the importance of particular facts for Gramsci's philology and Foucault's emphasis on the geography of micropowers underline how geography, seen as a dynamic concept of the territory and places, makes up the necessary condition for understanding institutions, forms of state, the management and circulation of power on a global level.[64] In many aspects, the continuous comparison of circumstances and particular localizations in Gramsci's epistemology, confirms the Foucauldian perception in which 'we live inside a set of relations that delineates emplacements sites which are irreducible to one another and absolutely not superimposable on one another'.[65]

Geography and Political Agency

Both the Southern Question and the Ford factory, seen as different scale examples of spatial strategies, re-evaluate the ability of each individual to influence or modify particular historical-territorial structures, while also identifying its oppressive or authoritarian nature. However, although Gramsci and Foucault are united by a similar anti-humanist orientation in rejecting transcendent solutions, such as the denial of the nature of science in the interpretation of history, or the Gramscian emphasis on a 'collective organism'[66] that substitutes the action of individuals, taking a closer look at their respective spatial policies causes a substantial divergence to emerge.

Unlike the Foucauldian concept of genealogy, the integral history wished for by Gramsci does not merely confine itself to recording the characteristics of an event, either confirming its uniqueness or its dispersion in the background of an era, but rather, in the aftermath of an actual disintegration, identifies and highlights those features that can then be useful in the establishment of an alternative historical block. Notwithstanding their shared interest in Foucault's 'capillary' and Gramsci's 'molecular' action processes, the French scholar's objective was to rebuild the process through which, at different periods in time, human beings have been brought to a state of subjection, undergoing the consequences of recurring 'dividing practices', both as actual segregation, even at a juridical level, and also as the parcellization of the social sciences: the result of this procedure, also evident in the accompanying in-depth and specific analysis, is a condemnation, fruit of high ethical standards, but which lacks any organizational or practical perspective. On the other contrary, for Gramsci, the so-called 'living philology', thanks to which we can subject the environmental conditions of a certain context to scrutiny, is realized by compartecipation, by 'com-passionality', in such a way that it prepares all those who will later take charge of social change, and move as a 'collective man'.[67]

The possibility of a social agent who takes charge of change, as already reported by Said, is the crux of the theoretical watershed between Foucault and Gramsci, which at least in part, limits Foucault to the social, but also geographical field of subaltern studies, where instead 'Gramsci, as Iain Chambers has argued recently, was instrumental in helping scholars rethink the understanding of historical, political, and cultural struggle by substituting the relationship between tradition and modernity with that of subaltern versus hegemonic parts of the world'.[68]

The above mentioned philology, as a methodological expression of the importance of particular facts considered as defined and specific individuality, is the basis for creating 'a critical and geographical rather than an encyclopedic or totalizing nominative or systematic terminology'. Said's comparison between Foucauldian and Gramscian procedures suggests, albeit still in rather vague terms, that 'Gramsci seems completely to have escaped the clutches of Hegelianism';[69] this emancipation from a Hegelian interpretation – the very thing that Foucault opposed – identified by the Palestinian scholar has been the subject of several recent studies on the philosophy of praxis; a philosophical approach that sees the recall of experiences as the means of connecting theoretical work to political praxis,[70] and which maintains close and constant relations with the geopolitical framework in which it operates.

Unlike Hegel's effect on Croce and the Hegelianism identified by Foucault as the origin of theological historicism, for Gramsci, Hegel's immanentism permeates philosophy and history, theory and practice, or rather, it produces powerful relations endowed with repercussions that are instantly political. Just as Kant before him, Foucault questions knowledge as the regime of truth, discussing in juridical terms, its legitimacy. On the contrary, Hegel[71] fuses thinking and doing, paving the way for a new and radical theorization of the relation between theory and practice. Starting from a Marxian reading of Hegel, Gramsci reconstructs a strategic concept, where history and politics coincide at the centre of a constitutive relation with praxis; this action in history is obviously developed in geographical space, a space in which crisis becomes the permanent condition of the international scene.[72]

In the wake of Marx, and his criticism of Feuerbach's use of abstract terms to place man within a materialistic universe, Gramsci addresses the question with a real – due to its great detail – concept and 'develops a process-based or relational understanding of the person. This person cannot be divorced from the natural world and the individual cannot be understood outside of specific socio-natural relations in particular places and particular times'.[73] According to Gramsci, the figure who can intervene politically must first of all be able to understand the relations that constitute the environment: 'the real philosopher is, and cannot be other than, the politician, the active man who modifies the environment, understanding by environment the ensemble of relations which each of us enters to take part in'.[74] But this 'real' philosopher is not a single individual, as shown in his analysis of the Ford factory, and does not correspond to the reassuring ideal of liberal humanism, whose subjectivism is in fact criticized, but is rather the result of a relation between monitoring action at a territorial level and collective intervention, promoting work initiatives by subordinate groups.

The philosophy of praxis operates within 'a worldliness and absolute earthliness of thought'[75] and it is language that provides the connection for grasping the coherence and interpenetration that exist between environment and society. This is not surprising, seeing that the concept of hegemony has its origins in linguistics, resulting from Gramsci's studies in this field,[76] and is inseparable from the socio-cultural and geographical theories of the famous early 20th century linguists, such as Ascoli, Meillet e Bartoli, who forged the tools for analyzing a spatial linguistics, where linguistic innovations occur at the same time, or as a consequence of social change. Using arguments on the translatability of languages and scientific languages, Gramsci correlates different languages from nations in a similar state of development, and establishes a comparison between them, explaining, after clarifying and discussing the parameters used for the comparison, why and in what way they are comparable. Translatability also refers to the possible relations between different cultures, as when Gramsci asks himself whether 'Machiavelli's essentially political language can be translated into economic terms and to which economic system it can be adapted'.[77] The process of translatability spreads, almost imperceptibly, from cultures to the environmental anthropological context, where translation does not necessarily imply the setting up of unequivocal linguistic correspondences between two distinct universes, but is instead a complex process, not only calling for language skills, but also requiring a profound knowledge of the cultural context of the territory under consideration and being translated, a concept first seen in Hegel's hypothesis on the translatability of French Jacobean thought into German idealism.

After due consideration of these elements, or rather, that of the position of the author of the translation, attention must be paid to the expectations and the characteristics of whosoever is on the receiving end, thereby determining the extent of innovation or confirmation of the thought existing before the translation, which has caused its rendering in a new context. Translation results as a process of negotiation between two cultural contexts of reference, and since neutral translations do not exist, the value of the entire procedure is measured through the sense and the approval that it can generate, always related to the specificity of the territories under consideration. From the philosophy of praxis, a Gramscian anti-deterministic and anti-materialistic, albeit brief, modus operandi emerges for translation, fruit of a radical reclamation of Hegel as an antidote, and of the phenomenological degeneration of Marxism which, paradoxically, Foucaultattributes to the selfsame Hegel . In particular, this last phase of translation appears as the dialectic epilogue to a circular path, in which we find a re-proposal of the very first queries about heterotopias: what strategies to adopt? How should one deal with different contexts, and what use should be made of the knowledge which we have? What statute should be accorded to the method used?

In this respect, the main difference, as mentioned above, has to do with Foucault's choice of referring to a subject who, in a Kantian perspective, represents an obstacle in the formation of social geography from an epistemological point of view: in fact, Kant[78] had not succeeded in applying his ideas on final causes to the field of geographical knowledge. Indeed, he complained that the organization of nature had no counterpart in the causality we know, and that this prevents the setting up of a form of geographical comprehension similar to a Newtonian style natural history. The metaphysics and the ethics theorized by the German philosopher were unable to find an appropriate field of application in geography, the reason why he gradually turned his attention to anthropology. And this is not all: Kant considered geography and history as being separate, in the sense that the former orders space, while the latter builds up a story in time and as a consequence, space and time are quite distinct in Kantian thought. According to Harvey, this sharp division between space and time not only reappears, but becomes radicalized in Foucault who had studied and translated Kantian anthropology, to the extent that in the *Herodote* interview, he talks to the geographers exclusively in terms of space and spatial structures. This is also reflected in the convergence between the application of Kant's theories and the result of Foucault's research: the laws on a local basis, the regional differentiation of a Kantian kind correspond, as we have seen, to a predilection for the microphysics of power and the contingent element, together with a reiterated rejection of the grand narrative.

According to Harvey, the problem lies in the fact that Kant's whole approach to geography and space is based on a concept of space and time inspired by Newton, in which they are seen as absolute entities, whereas valid arguments exist,[79] beyond the scope of this present article, for considering space as a relative and absolute entity at the same time, and consequently, geography and history are viewed as two sides of the same coin. Since Foucault claims to consider a geography as a 'condizione di possibilità' [condition of possibility] for other forms of knowledge, heterotopias still remain anchored to Newton's vision of space, and as such, are situated at the threshold between the possibility and impossibility that a spatial change may effectively come to pass. This, however, does not undermine the effectiveness of the aforesaid critique in any way: in fact, Foucault's consideration of space goes hand in hand with deconstruction, which underlines how identity, playing on a principle of inclusion/exclusion, is not only created on the basis of the definition of the Other, but also tends to impose a cage on equals, from which it is then very difficult to escape.[80] This radical criticism 'of the illusions and contradictions in which contemporary thought seems to shut itself up' is where Foucault finds an analysis and reporting system in Gramscian hegemony that interfaces with his own, resulting complementary in the praxis that follows the analysis, as confirmed by the copresence of two, among many others, verifiable critical approaches- in the first instance, Said and also in modern *subaltern studies*. In all these spatial contexts, hegemony is the power 'to determine that which can be said, or proved. If so, hegemony exercises power through the constitution of a world, through making experience of that world possible and thus creating some language games and silencing others':[81] thence, for hegemony read heterotopia.

2 'AN OCCULT GEOMETRY OF CAPITAL': HETEROTOPIA, HISTORY AND HYPERMODERNISM IN IAIN SINCLAIR'S LITERARY GEOGRAPHY

Tom Bristow

Geography wrests history from the cult of necessity in order to stress the irreducibility of contingency. It wrests it from the cult of origins in order to affirm the power of a 'milieu'.

Deleuze and Guattari
What is Philosophy?[1]

Every historical era ... is multi-temporal, simultaneously drawing from the obsolete, the contemporary and the future.

Michel Serres and Bruno Latour
Conversations on Science, Culture, and Time[2]

2009: The State of London

The Olympics is one of the hallmark events of the global capitalist culture industry alongside world fairs and city expositions. Performing an economic role similar to heavy industry in the previous century, they raise political and moral questions with respect to resource allocation, scales of investment, environmental degradation and affordable housing. The Olympic Movement of the modern era is a globalized institution; its mega-events tendered as catalysts for urban regeneration policies with profound socio-cultural effects.[3] Public spaces are one of the discursive effects through which power works, and in the context of environmental impact on sites transformed to house the games, heritage is a contentious issue.[4] This chapter considers an artistic response to the politics of post-industrial England, particularly in the context of urban development and nation building; it pays particular attention to the issue of social cohesion as disclosed during the London Olympic bid. The private use of space in the Lower Lea Valley during this cultural moment is indicative of the market driven politics of East London throughout and after the Thatcher years: new homogenous high security housing developments disconnected from local history, diversity and complexity; decline in use of and access to shared green spaces. The cultural fabric is being ripped apart rather than sewn together. A counterpoint to this example of an institutionalized binding process is Sinclair's psychogeography of London that is alert to rich accounts of geography textured by difference and open to the dynamics of national and global politics. These histories are brought to life by animating a range of interpretive grids of human culture; literary, geographic, and economic filters and frames generate clear views on life within specific spaces. Some of these spaces are heterotopias – belied power geometries of culture and communication (discourse, exchanges) – which reclaim slower processes of radical energy exchange as resistance to contemporary political cultures that figure the city as cosmopolitan souvenir.[5]

The urban imaginary can explore various layers of history to expose the fragile construction of our shallow heritage formations, our surface culture. Andreas Huyssen has identified a compulsion within the city's imaginary's reach: '[it] may well put different things in one place: memories of what there was before, imagined alternatives to what is present'.[6] Iain Sinclair's poetry, novels, and literary criticism trade in specifics rather than universals; in an insightful attention to the capital of England through micro-observations connected in curious ways, Sinclair attempts to undermine authoritative voices. This manifests on two fronts in his London texts. First it is directed towards the map of power, as his prose demonstrates explicitly; the second faces in another direction towards intangible, invisible, unconscious drives (fears, anger, pleasure, hopes) and archaic energies, both dislocated from the hegemonic power and not constituted by structural relations. Sinclair's literary cartography is animated by spatially and temporally submerged energy; operating beyond the time-space compression of global capital it resists narratives that dematerialize and conflate properties and qualities of spaces.[7] Michael Moorcock has understood this aesthetic in terms appropriate to Angela Carter, Philip K. Dick and J. G. Ballard – all 'extra mural romantics', with Sinclair's poetic rendering of the past figured as a force keeping it alive 'still dripping... stinking... kicking'.[8] Heterotopia, for Sinclair, is akin to a third space, a poetic site of resistance – as exemplified by the motif of Nicholas Hawksmoor posited within Sinclair's documentary fiction (below) – a politicized space that puts a halt to mediated data (culture) slipping into law (history). Such grass roots and pedestrian resistance disclose spaces that are akin to those 'dimly lit, opaque, deliberately hidden, saturated with memories, that echo with lost words and the cracked sounds of pleasure and enjoyment'.[9] This complex semiotic fold magnetizes mind to world; its intellectual orientation and emotional comportment to space foregrounds dynamic histories which resist normative cultural enclosures as instanced by late capitalist post-industrialism.

London: Capital as Achilles Heel

The global city as platform for international economic success is associated with post-cold war Olympiads moving beyond cities' reputations as 'harbinger[s] of social decay and economic depression'. Olympic cities are multilayered and offer changing perceptions of the city as a source of economic and cultural dynamism rather than as a symbol of social decline and decay.[10]

> East London is a relatively deprived region of the city where traditional manufacturing industries and an extensive docklands area experienced closure and de-industrialisation in the 1970s and 1980s. Regeneration has occurred in specific spaces and places over the past twenty-five years. This process of decline and renewal has been matched by the fluidity of the area's population, with movement inwards and outwards creating a uniquely multinational and multi-ethnic population. East London experiences a heady mix of social inclusion and exclusion and poverty and wealth. It was these conditions that the London Olympic bid was designed to address – an ambitious programme of urban renewal backed by both central government and the City's Mayor.[11]

During the manufacturing era, parts of East London outside the control of the former London City Council, were subject to polluting industries, garbage disposal, landfill, and car-breaking. The 2012 Games and associated new housing were built on these already ruined sites. As Pointer and MacRury identify, following manufacturing losses the region has suffered catastrophic decline in areas as broad as paper and cardboard manufacturing, cement, oil refineries, and car manufacturing. London draws resources from capital investments to labour skills, from Europe and beyond; yet it fails to economically (and culturally) produce itself internally. For Sinclair this lack of capacity to self-generation suggests a weakness in our wider culture: our historical consciousness and a breakdown in the tradition of social critique and collectivism.

Findings in cultural studies suggest that the rational, economic gloss of governments mobilizing GDP figures, only adumbrates the imperative for social cohesion; it does not flesh it out. As things are being lost in historical, geographic and cultural terms, Sinclair is increasingly interested in 'the way that we assemble evidence'; owing to his critical eye that is attuned to market information and data often presented in relation with degrees of ability to generate culture.[12] I read Sinclair's sensitivity to heterotopia as an ability to incorporate the irony of what Baudrillard has called 'surface accounting', where 'an interest in surface does not mean a disinterest in the wider systems in which a "thing" is entangled (be they systems of production, inhabitation, valuation, or dissolution)'.[13] As such, it attempts to reanimate space and connect to deep energies from within the enclosure of hegemonic capital. In human geographic terms, 'regeneration' is not viewed as the process of making places meaningful, but the site of the loss of generations of human memory incalculable in the context of rapid development. For Sinclair, this foregrounds the need to contemplate the possibilities of 'reverse archaeology' – to record in a landscape traces of life that have inhabited the space before new developments are built.

Reverse Archaeology

Sinclair's representations of the Olympic development site in the Lower Lea Valley, London Borough of Hackney, do not subjugate their evident spatiality to an aesthetic trope or to a thematic contour. Heterotopic sites in Sinclair's fiction and journalism collectively forward a poetic geography that challenges the restrictive zone of a rationalized geography of 'literal, forbidden or permitted meaning'.[14] This counter-culture of experience is partly occult, partly polemic.

Two articles in the *London Review of Books* inquire into Olympian capitalist spoils and reread Sinclair's contemporary, the London scholar and chief archivist of historical anecdotes, Peter Ackroyd. Ackroyd and Sinclair are two of Britain's capital's literary giants with significant purchase on the London imaginary. In part a reaction to Ackroyd's conservatism, Sinclair's cybernetic and regenerative 'reverse archaeology' critiques history that legitimates the present. Such resistance writes against the grain of spatial heritage; spaces that might be written up into accounts of lifelihood always-already mediated and determined by pre-existing history. As method, it probes beyond the inscription of life-worlds oriented to an idea of the nation as one that has evolved through time; it moves beyond 'the tsunami of speculative capital ... and 2012 game-show rabies'[15] to uncover false passive inevitability. Ackroyd has posited a resistance to historical knowledge in the figure of the astronomer in *First Light* (1989). As one of a party of eccentrics that discover a Neolithic grave, Ackroyd's character places emphasis on the linearity of time when discussing a collective approach to read deeper significance within things: 'You can never go back ... Signals sent into the past would be killed by their own echoes. You can only do one thing. You can send signals into the future'.[16] In direct opposition, one assumes, is a mind open to the unfinished disclosures and unfolding of energies and stories, once located in the past. Sinclair: 'This is the loss we fear most: the contemplative solitude of the water margin, its accumulation of voices. Rivers and canals flow through us, changing and not changing, catching the rays of the rising sun and the transit of clouds'.[17] The latter voice, constructed while moving along the towpath from Camden Lock to Victoria Park, denotes a resistance to fixed parameters (margins) and metanarratives (a homogenized synthesis of an accumulation of voices). I claim this voice as a compliment to Foucault's sense of heterotopia.

Sinclair's cultural critique learns from literary and cultural modernism and its associative project of literary geography that endeavours to understand what David Harvey has detected as 'time-space compression', a view of the world as one that seems to 'collapse inwards upon us'. For Harvey, the collapse signals the change to our central value system, which 'is dematerialized': 'shifting, time horizons are collapsing, and it is hard to tell exactly what space we are in when it comes to assessing causes and effects, meanings or values'. It is not precisely heterotopic but it gestures toward some of the confusion evident in such discordant and rich arrays. Moreover, time-space compression, which manifests in 'the interweaving of simulacra in daily life', Harvey states, 'brings together different worlds (of commodities) in the same space and time'. Most pertinent to Sinclair, is that 'it does so in such a way as to conceal almost perfectly any trace of origin, of the labour processes that produced them, or of the social relations implicated in their production'.[18] To bring these processes in to view is to keep the space alive and to discover its meaning; perhaps this is to translate it into a historical materialist heterotopia. With this in mind, we can read Sinclair's project as one engaged in the possibility of opening up this inwardness to foreground traces, social relations, and by implication, the processes of capital and other suppressed energies underneath the surface gloss of the repressive grid of social control. Sinclair appears to convert surface cultures to heterotopic environments, deep with time; arguing that traces can produce new meanings, precisely those that are required during London's regeneration and its crisis of representation within the context of heritage culture and surface histories.

Knowledge Economies and Enclosures

The Olympic development site was first secured in the public imagination once it was given spatial definition by a blue perimeter fence, fifteen feet tall, enclosing 500 acres of previously open and accessible land. Secured in some places with a 5,000-volt security fence (razor wire), the boundary marker was a significant symbol in Sinclair's prose of the time.[19] As an example of just how quickly the geography can change, or an exemplary manifestation of 'termite activity, the neurotic compulsion to enclose and alienate',[20] this material circumference – complete with 900 CCTV cameras – also acted as canvas. During the lead up to the Games, posters of imagined Olympic events drawn by children from local primary schools were pasted onto the walls. Later these were taken down as the public they generated posed a security threat, it was deemed. Local artworks by children were replaced with polished images of the anticipated events, consistently branded with the iconography of the London 2012 marketing logo. An example of mediated data slipping into a force of power, controlling the space.

This is how Sinclair reads the heritage industry surrounding the Olympic developments. For Sinclair the authoritative virtual images of the games in the future, which have won the battle of the perimeter wall, are the énoncés of the legislators of occult capital or the power-geometry sometimes registered by the phrase 'the heritage industry'. Foucault's definition of 'énoncés' as 'a general system of the formation and transformation of statements' seems apposite here. Traditionally, 'énoncés' denote a technical or formal statement that belongs to discursive formation. However, Foucault elevates the term from discourse to system: with a focus toward power relations, the system signifies an economy of knowledge, the language and medium constituting a discourse.[21] At the time of this exchange – or transition – from diverse, organic self-expression and determination by local culture to the imposition and control of hegemonic counter-narratives, the latter was undoubtedly conceived as an unstoppable force set to deliver a significant project without delay – 'the 17-day corporate extravaganza ... to which we are all so deeply mortgaged'.[22] Moreover, its aesthetic form – a complex boundary zone terminating life-flows (present restriction) while indicating the capacity for a potential reality (future orientation) – changed the direction of propagation that suggests inescapability, or in Sinclair's words, 'it had happened, it was happening, and it described the future we are now experiencing'.[23]

Sinclair's project speaks directly to England's first novelist, Daniel Defoe. A novelist who has been imagined as one figure within a line of dissenters that give rise to a critique of Protestantism and Capitalism, leading from the Restoration, which signifies a failed revolution and cultural enclosure – as significant as the land closures, the reduction of the commons – in English life. The critical geographical context might amplify this loss of respect for communal space as a breakdown in place-making experiences. The act of enclosure that is symbolized by the boundary zone around the Olympic site – demarcating private property under high-security surveillance – 'justifies itself', Sinclair argues, 'by exploiting temporary fences to use as masking screens, notice boards for sponsors' boasts, assertions of a bright, computer-generated future'. This is corporate data, not cultural expression. The ring exhibits 'on message' exhibitions of 'sanctioned street art', part of a machine that makes 'clever move[s that] pre-empt the attentions of spray-can subversives, class warriors, animal liberationists and wannabe Banksies'. In this roller coaster of vehement criticism shot through with sardonic humour, Sinclair sought to find the energies of 'community' and the contestations of space by local groups and activists.[24] What he discovered, it appears, was a collective body politics in concord with the new post-1980s political cybernetics, defined as: 'the interaction of autonomous political actors and subgroups, and the practical and reflexive consciousness of the subjects who produce and reproduce the structure of a political community'.[25] The new, or second, cybernetics problematizes Alfred North Whitehead's metaphysical concept of 'subjective aim' within all natural things by leading towards an information technology perspective that remains informed by living systems' biological organization. He is thus in line with David Harvey's understanding of New Right economic policies masked by postmodernism's 'cultural mystification or camouflage' (see Figure 2.1). The heterotopia of the perimeter fence suggests that the excitement of glamour or saturation in imagery and stress on individualism and self expression of many western inner cities acts as social control, they are 'consciously deployed to pacify restless or discontented elements in a population' – as evidenced by shopping malls, office towers and the act of security that is the Olympic Park perimeter fence normalizing and appropriating attempts at subterfuge.[26]

Figure 2.1: The Queen Elizabeth Olympic Park. © Martin Pettit

The cultural mark of the security zone is, for Sinclair, 'the hinterland between the virtual and the actual'. These virtual images of the geography in the future, either during the Games (pictures of athletes in various park locations yet to be built), or the bourgeois unimaginative pastoral dwelling places of the post-Games landscapes lying in the wake of Europe's highest levels of private security, are gestural facades of heritage culture fly-posted on the perimeter fence in an act of occupying legible space that signifies selective memory or strategic capital history.[27] The shift from heterogenous primary school to official homogenous corporate brand icons symbolizes the political and cultural climate of the times. Sinclair describes how:

> [L]ong established businesses closed down, travelers expelled from edgeland settlements, allotment holders turned out – there were meetings, protests, consultations. As soon as the Olympic Park was enclosed, and therefore defined, loss quantified, the fence around the site became a symbol for opposition and the focus for discussion groups.[28]

The perimeter fence is supercharged for Sinclair; it acts in a similar way to Stuart Elden's remark on Foucault's conception of the transformation of statements: '[where] the discursive formulation of a subject also acts as a limit'.[29] Historically we impose form when there is none; we invent nouns for transitional objects as mnemonics that aid our reflective and critical manipulation of things. The idea of a 'subject' is limited when read in terms of larger fields of power, distributed agencies and global ecologies. Somewhat distinct from Foucault's inquiry into how an institution or a set of *a priori* conceptions arose, however, Sinclair starts with the present as *underdetermined* and *evitable*. Quite the reverse of the energy of capitalist development and heritage construction, Sinclair's connection to English cultural formations undermines (an Ackroydian) model of history that offers nothing short of a legitimation of the present.

Upriver and Game-Show Rabies

Geography is exercised by the problem of defining regions. Sinclair's local mapping understands that drawing a line around a place does not enable shared values between developers and local inhabitants to flourish. It creates stasis and standoff distance. Conversely, his critique is programmed to connect to cultural histories, to emphasize flow and to illuminate people's life ways and their ability to interact with the physical environment and its stratified history. While the future of the area is undecided, being subject to new ideas and new market forces, Sinclair states that the short-term project for literary geographers of the Olympic site: 'is to quantify loss'. While people were expelled from the area, and most ironically, swimming pools, cycle tracks, and football pitches were closed, Sinclair posited a progressive position with respect to the fact that 'the landscape is so powerful and vertical in its resonances' that local artists, rather than using 'cosmeticized dissident voices in reacting to the [suppression of information and the terms of the legacy at present] are forming a new spirit: these virtual images will be *overwritten by calligraphy we wish to make*'. Loss and legacy are keywords to Sinclair's output during this period; they act as a gloss to a rich layer of the cultural palimpsest; a key to darker materials and sources, which I view as elements within his heterotopic space.

It is noteworthy that Sinclair has spoken of an 'inclination towards apocalyptic conspiracy theories, palimpsests of gangsterism, bad politics' in his work; and he has clarified an awareness of how 'cynicism can atrophy into lazy sentimentalism'. Centrally, for the cybernetic Sinclair, the problem of origin and nature of matter has been overlooked by the quest for the origin of order. Rather than focus on energy, it appears that western societies are fixated with forms; vibrating fields of interaction over time are lost for an inclination towards the photograph, the simple two-dimensional shot of life. Sinclair's reverse archaeology, a negation of the quest for order – sharpened by the polemic of the attacks on Ackroydian history configured as continuity, legitimation, and inevitability – restates Gregory Bateson's notion of explanation as the move toward form rather than the move towards substance. Form, and its cousin, identity, suggest that which is solid; substance suggests openness to reformulation. Both Bateson and Sinclair inquire into the experience of data, the raw facts of life, practice and behaviour; existence preceding essence.

Sinclair's historical outlook intuitively recoils from the attempt to build a simulacrum of the phenomenal universe in static images and words owing to the false prophet of positivist, causal explanation and total history lurking in proximity to large-scale rhetoric (ignorant and blind to probability, as with the futuristic marketing images on the fence). For example, his concern with Ackroyd's formulation, 'sacred river', as a metaphor for *continuity* is keenly associated with Ackroyd's celebration of the developments for the Games in terms of realized potential i.e. London fulfilling itself, manifesting into an idealized (preconceived?) form. This is not a culture generating itself through change. Such teleological construction is read as a product (the verbal transformation and censorship of the phenomena) in Sinclair's reading of Ackroyd's censorship and critical confusion when mapping the river in his London works. Sinclair's criticism details Ackroyd's shift to metaphor from the presentation of raw data on the Thames – its length, velocity of current etc. It is a move made quite rapidly:

> So that the two tendencies, the empirical and the poetic, coexist, informing and challenging each other, striking examples found to confirm flights of fancy. And all the time [Ackroyd] is walking, from limestone causeway to salt marshes, but keeping the accidents and epiphanies of these private excursions out of his narrative.[30]

Embellishment is kept to a minimum; personal experience is suppressed. Ackroyd's abstraction claims the objective goal and strategy of induction: to enumerate a number of facts for the purpose of a general statement, which is a total history that mimics the form of an open (or unfinished) heterotopia. Sinclair instinctively examines how the rules of this transformation (from empirical to poetic) and the differences in coding between natural phenomena, message phenomena, and words instances Bateson's outline argument: the problem of mediated data slipping into law.[31] To remind ourselves: cybernetics stems from the Greek root, kybernétés (the steersman or pilot of a boat) from which we also derive 'government', which indicate systems that interact with themselves and produce themselves from materials of their own making.[32] We are now attuned to the import here. Sinclair quotes Ackroyd: "'The journey towards the source is the journey backwards, away from human history'";[33] thus, to walk towards the source of the river Thames in the manner that Sinclair does (when reading Ackroyd's text), is to embark on a journey to uncover hidden potential, lost narratives, and raw data while resisting homogeneity and hegemonic geometry – it instances non-causal self-regulation. To light out for the territory.

Sinclair's challenge to Ackroyd's conception of the river as a deity ('sacred river') begins with a 'series of expeditions along the *permitted* riverpath *from mouth to source*', a significant reversal of direction coupled to implied restriction (permission) enables something that Sinclair calls 'a more cynical view', which can read the 'organic entity forever renewing itself from the darkest sources'.[34] This is central to Sinclair's alternative version of London, and his nuanced heterotopia. It follows an observation: 'on the rough lawn in front of the improved Haggerston flats, there is a chart, behind misted glass, in a wooden cabinet designed for community notices: a premature map of the Olympic legacy'. Inspired by this 'indecipherable' text, Sinclair decides to plunge himself into the river's present history with one of Ackroyd's books in his hands.[35] The pre-Olympic dwelling of Adelaide Wharf residences are at the North-west end of Haggerston Park; Haggerston, we should note, not only represents one example of this failing, expensive, 'spanking new canalside development in loudly upbeat colours', but the site itself is a rich signifier, a collision of past and present. Remarkable as it is, like the Olympics, an enclosed park, and like parts of the Olympic site, Haggerston is a polluted area lightly covered with a thin gloss of surface acceptability, or more literally, materials covering up the recent past. A maligned attempt by the capital gloss to suppress dark energy in the park resonates with Sinclair's interest in the depth of London's misery, and parallels the contemporary critique of polished surfaces lacking integration with their history: 'polluted acres of the Imperial Gas Light and Coke Company were recreated, after war and bomb damage, as Haggerston Park'.[36] Readers are mindful of the cultural constructions on the perimeter fence. In this *LRB* article, the flat-dwellers in a cocoon of their own making are married to twice-exiled Polish builders (self-imposed movement from home; displacement outside the security of the workplace). Economic diaspora meets historic and contemporary toxic capitalist colonization of space. A panoramic view of this space includes rough sleepers making what they can of the park and its padlocked gates. Images resonate with anecdotes of suppression littering Sinclair's article: disconnection, silence, pressures of capital in a single literary compression. This situated knowledge sustains a critical distance from the (permitted, hegemomic) forces that envelope an emergent stratum of living systems most notably social relations, human subjectivity, and environmental concerns.[37]

This ecology comes to the fore in Sinclair's critical evaluation of Ackroyd's depiction of the Thames as 'a mirror for national identity' and 'constant in history'. The river as 'unifying metaphor' reminds him of the narrative offered by Henrietta Marshall's, *Our Island Story: A History of England for Boys and Girls* (1905), memorable for its illustrations rather than prose, which is 'broken down into Jamie-Oliver-sized portions suitable for juvenile digestion'.[38] Sinclair explicitly marks his venture as 'a reverse Ackroyd walk', archaeology over history, which could also be understood as resistance to the eastward flow of capital if it were not to metamorphose into a pleasurable stride (with Ackroyd's text in hand) that writes Sinclair's private experience into the landscape adjacent to a portable text. Sinclair is 'amused by the sight of a rowing eight so preoccupied by their furious activity that they wedged themselves in a thin channel cut through ice'. These figures metonymically signify Ackroyd's teleology; their 'oars scraping plaintively and impotently' as if their pursuit fails to make purchase on the deeper current that is primary energy in Sinclair's darker, critical outlook. Sinclair: 'One of the distinguishing features of Ackroyd's Thames is recurrence; landscape is revised, personages come and go, the nature of the river never changes', we are told. Recursion in Sinclair is more complex as we shall find.

As a conservative myth for Sinclair, recurrence links Ackroyd to the spoils of regeneration and heritage culture. Ackroyd's marketed progression is ironically posited within a historic period by Sinclair's reflective narrative voice, as near-future proleptic consciousness looking back on the present. It reads like a still image of a heterotopia with wide-ranging cultural signposts collapsing under its own centre of gravity – association:

> The reimagining of downriver stretches of the Thames was not limited to East Greenwich: fantasy settlements were imposed on vacant brownfield sites along the floodplain in Essex and Kent. Every act of demolition, every fresh-minted estate, required a recalibrating of history: as a hospital or asylum vanishes, we thirst for stories of Queen Elizabeth I at Tilbury or Pocahontas coming ashore, in her dying fever, at Gravesend. The documented records of the lives of those unfortunates shipped out to cholera hospitals on Dartford Marshes, or secure madhouses in the slipstream of the M25, can be dumped in the skip. Politicised history is a panacea, comforting the bereft, treating us, again and again, to the same consoling fables. Laminated boards appearing around loudly hyped newt reservations or permitted greenways, punch and partial summaries of an approved narrative of the past, found their equivalent in the 2005 reissue of *Our Island Story* by the right-wing think-tank Civitas. John Clare, the education editor of the *Daily Telegraph*, appealed to his readers for donations to support this project. 'They responded by sending an astonishing £25,000.' There were messages of endorsement from Lady Antonia Fraser and the feisty historian Andrew Roberts; the *Economist* saluted the new edition as 'impeccably postmodern'; 5000 free copies were distributed to schools, a Trojan horse for early indoctrination in traditional values that would be reinforced by emphatic TV explainers vamping through the palaces and bedchambers of the Tudors. Supporting copy, put out by Civitas, warned that 'people, including politicians of all parties, are worried by the failure of many young people now to engage with the institutions of the free society they live in.' Denied access to the pieties of *Our Island Story*, a generation of misguided eco-protesters and climate-camp activists might find themselves on the wrong island, at Kingsnorth Power Station on Grain, kettled between Medway and Thames, waiting to be filmed, fingerprinted and battered by the successors, as Akroyd might see it, of Richard II's Blackheath enforcers. There is always a TSG Metropolitan Police Force:Territorial Support Group (TSG) presence, only the uniforms change. Now identification numbers vanish from shoulders and medics carry extendible batons.[39]

John Clare's alarmed 'environmental-Romantic' walk to Northampton disclosing industrial changes to the landscape is merged into a brief acknowledgement of Samuel Beckett's Ham from *Endgame* recast as Harold Pinter in a staged and pre-recorded Nobel Prize speech on the distinction between fact and fiction (Fraser),[40] is swiftly followed by the analogy between the London Metropolitan Police Force's Territorial Support Group (TSG) and the English Peasants' Revolt of 1381. These imaginative collisions emphasize change and difference (the conditions of the present) – not continuity, not similarity. Sinclair is a celebrant of the evitable and of alternative ideas; new associations give life to each element and extend their meaning backwards through time while opening up new intersections on the hyper-modernist matrices of interpretation. Ackroyd's progressive historicism, which is reflected in his publication strategy – to have one text 'leaking' into another, as Sinclair suggests, is to always press forwards 'not backwards into a revelation from some unnoticed aspect of the past'. This rhetorical strategy figured as offspring from methodology 'defines the precise moment' for Sinclair, 'at which locality becomes location'[41] – when the dynamics and energy of space, infused with various power-geometries throughout time shift from lived experience to abstracted context (from data to law). From thing to world, verb to noun, process to stasis, heterogeneity to homogeneity, heterotopia to tradition and singularity, life to commodity. How can commodities regenerate themselves?

Locality as location, thus conceived more broadly, signals post-modern inauthentic heritage closely aligned in Sinclair's prose to his critique of the marketing strategies of the Olympics. Ackroyd's prose, operating in the 'state of limbo [where] past and present [are] interwoven', structures 'recurrences and interconnections' to disguise a 'lack of content'. This impotent void is an ideological stance to Sinclair. In praise of 'shopping areas and apartments' in his reading of St Katharine's Dock by the Tower of London, Ackroyd's historical view suggests that "'the neighbourhood of the river is recovering its ancient exuberance and energy, and is reverting to its existence before the residents and houses were displaced by the building of the docks in the nineteenth century'". Historicism here is of great interest and yet it is progressing towards form and inevitability; this law is accompanied by a vision of the river as the highway of the nation: a principle urban resource.[42] Regeneration from this perspective is viewed as resettlement of a prior state, "'it has not greatly changed in the last 2000 yrs'"; Ackroyd's voice framed by Sinclair in the context of 2012, is spatially arranged in proximity to Sinclair's polemical outburst on pro-development perspectives coupled to metaphors of constancy as a 'megalomaniac right-wing fantasy'. There is a tension between cultural asphyxiation of progress on one hand, and associative history reaching backwards on the other. In a final move to this chapter, I implicitly revisit this clash of ideologies while considering how the historic built environment both performs and lends itself to regenerative cultural practice in the context of a literary critique of normalized power.

Energy Versus Fields of Force

Both Sinclair and Ackroyd have investigated Nicholas Hawksmoor's symbolic importance to London's iconography. Hawksmoor's churches exist inside and out of time; they are engaged with their eighteenth-century history and continue to impress upon today's London's writers. I consider two of three Upper Wapping churches begun in 1714, namely, St Anne's, Limehouse, and Christchurch, Spitalfields, and I will introduce how these are conceived within a context of regeneration vis-à-vis Sinclair's politicized 2012.[43]

Hawksmoor's sources for these churches were threefold: the late seventeenth-century interest in Early Christian buildings considered in the light of the Reformation emphasis on primitive observances; Sir Christopher Wren's historical interest in pre-Gothic and pre-Classical styles; the influence of the 'ideal' centralized plan from the Renaissance onwards.[44] It is worth stating that Sinclair's prose poems have read St Anne's forecourt area as one set aside for sacrifice and fire ceremonies: not only the site of rituals of purification, but a space that 'reaches back, through an early-Christian sense of protected dwelling-place and stable, to the church as host'.[45] Sinclair's project to recover lost energy counteracts the present where 'These facts fade. The big traffic slams by. A work ethic buries ancient descriptions'.[46] Furthermore, particularly sensitive to materials and place, Hawksmoor's involvement in St Luke's obelisk at Bunhill Fields, not only plugs-in to a fascinating, if somewhat morbid literary heritage – the burial place of Blake, Defoe and Bunyan and the place of Milton's death – but operates as an energy field of focus for these multiple histories, a 'sequence of heated incisions through the membranous time-layer' as Sinclair has it.[47]

In the case of Hawksmoor, each site is a medium of its history and architectonic channelling of energy beyond its individual locus; each constitutes part of a subversive semantic network that can offer new navigations of terrain beyond geometric surface accounting and can rewrite the city outwith the grammar of heritage culture. These are important heterotopies in and of themselves. Furthermore, as an example of this rich semiotic spatial field, Dave McKean's maps for Sinclair's volume of poetry, *Lud Heat*, represents 'lines of influence' across Hawksmoor's churches from the most westerly (St George, Bloomsbury), to the most easterly (St Anne's, Limehouse), to spatialize 'the invisible rods of force' or dynamic energy running through London.[48] This energy is not only represented symbolically in the architecture; it is harnessed in each building's conversation with its place, and with the history of its aesthetic form and its own interpretation of that history (i.e. regeneration). Sinclair:

> [F]rom what is known of Hawksmoor it is possible to imagine that he did work a code into the buildings, knowingly or unknowingly, templates of meaning, bands of *continuing ritual*. The building should be a Temple, an active place, a high metaphor. The buildings taken together, knotted across the city, yield a further word.[49]

'A further word' denotes ongoing generation and processual semantics. The paramount context for the silent, elliptical hieroglyphics of Hawksmoor's (Debordian) distinct architectonic poetics is: murder.[50] Conversely, when speaking of the poet, Aidan Dunn, and his epic poem locating King's Cross as the spiritual centre of London and St Pancras Old Church as the epicentre for spiritual rebirth (*Vale Royal*, 1995), Sinclair cites Ackroyd's supporting letter to Dunn's publisher for jacket copy: 'an extraordinary sense of the past ... one of those people along with Blake, Chatterton and others, who are like a divining rod for history'; Sinclair counters this linearity by writing of Dunn's diachronic aesthetic as a 'helic structure', the reversal of 'the city's entropic energy field'; somewhere between the New Physics and orthodox fantasy.[51]

Sinclair continually revisits death and regeneration in an attempt to marry the 'unacknowledged magnetism' to 'unresting London.'[52] Like the filmmaker Patrick Keiller (drawing on Defoe's early psychogeographic vision),[53] Sinclair reads this combination of the surfacing of dark energy and suppression as part of 'the problem of Britain'.[54] As part of an ongoing history of recording energy in a landscape, Sinclair's Hawskmoor instances a wide and thick sense of surface and of time. As such, it is an exemplary model of the Bakhtinian chronotope crystallizing the intrinsic connectedness of temporal and spatial relations:

> In the literary artistic chronotope, spatial and temporal indicators are fused into one carefully thought-out, concrete whole. Time, as it were, thickens, takes on flesh, becomes artistically visible; likewise, space becomes charged and responsive to the movements of time, plot and history.[55]

The chronologic perspective provides human geographers with insight into Sinclair's spatial 'sense-making', which is remotely heterotopic in that it operates on two levels. Firstly, a critique of space and capitalism where heritage is 'out there', both as commodities or 'inventions', and 'maps of meaning'. Furthermore, Sinclair's psychogeographic impulse is charged by a British social critique to understand the cultural material out of which broad moral and historical systems are made, reproduced and turned into facts, signposts, codes and context for cultural memory.[56] Secondly, the ethnographic spatial imaginary of private experience; this is an aesthetic committed to observation and dedicated to motion, invention and transformation of geography, and it is deeply respectful of individual phenomena and the cultural archive. These two energized dispositions suggest that culture is neither free-floating nor mere icing to the capitalist economic base.

In Sinclair, Hawksmoor is figured within a complex geometry of reflections and critiques of artists, architecture and processes – he renders the city as a meta-spatial discourse. This deserves a full study in its own right; however, for brevity some aspects are synthesized here as a means to illustrate Sinclair's orchestration of epistemic chaos as a means to unearth complex history. I am selecting only two sources to highlight part of the qualitative dimension that Sinclair's use of Hawksmoor generates and proliferates within his oeuvre. First, the authority on Hawksmoor, Kerry Downes, has noted that St Anne's interior wall is unique in that it is 'an organic medium with its own life'; this building echoes Christchurch's complex interiors that evoke the whole building's 'freedom of its sources'.[57] This internal vitality is continued in the method of structural development instanced by the pyramidal pinnacles on St Anne's towers, which we know are 'the last of all in [Hawksmoor's] sequence of drawings', much like Bloomsbury's St George's unique and obscure stepped pyramid on columns (that forms the base for the steeple) was not designed before building commenced.[58] These all suggest that Hawksmoor resonates with Sinclair – inquisitive with respect to how one can respond to emerging forms, to be 'wary of closure', to follow 'flow ... momentum ... accumulations', while also indicating the larger discourse of enclosure: St Anne's 'padlocked gates' suggest 'no sanctuary' from enforced security, Sinclair remarks.[59]

Second, the appropriation of local artists working with Hawksmoor's buildings embedded into Sinclair's work (without irony) locates a precise aesthetic of: 'fanciful arrangements ... [which] take on meaning' as he remarks on Gavin Jones – 'painter, sculptor, earthmover, outlaw ecologist'.[60] Jones places his works of art *inside* St Anne's, Limehouse, transforming the space from the museological sense (where the heterotopic site encloses objects from all times and styles) to '[an] intense displacement of energy'. Despite closing off the building to visitors, Jones' 'work was unaffected by the fact that it could not actually be seen'.[61] The reader/ visitor is encouraged to attune to the arrangement of work within a work; meta-poetics of framing positioned within Sinclair's work in turn constructing a new semantic field that runs beyond the enclosing space. By incorporating local, innovative 'off-radar' artists, Sinclair illustrates how place (textual and geographic) operates and shelters 'entanglements and configurations of multiple trajectories, multiple histories'.[62]

Sinclair's prose operates within an intertextual site that 'is not word perfect, [for] its gematria is not made with full consciousness'. It is associative and cumulative; open to new words and new values that rise through new associations. Sinclair's explicit invocation of the mystic value of letters in scripture revealed by Cabbalistic methods of interpretation (gematria) belies the fact that Hawksmoor's dynamism is true to the contingent and intuitive. Sinclair has noted this contradiction by writing that Hawksmoor's coded non-random design enables 'accidents [to] occur'.[63] Due to this site's very capacity for newness, as a space for Jones to place his works inside the 'chill, baroque interior' of St Anne's, it is to Sinclair an 'ideal setting' in which site and artefact enable 'a chance for something unexpected to develop from the collision'. Again, non-prescriptive energies emerging from interactions with place. In terms of a post-Foucauldian heterotopic state, potential events remain alive with possibility within an open historic framework. This rudimentary synthesis can help to uncover Sinclair's histrionic mode i.e. his playful aesthetics and the move towards a history of the present via reverse archaeology.[64]

In *Lights Out for the Territory* (1997), Sinclair glances up at One Canada Square (the Canary Wharf Tower), the ultimate symbol of failed late 1980s private sector investment in Thatcherite Britain, 1990s deregulated economic turnaround, and 2000s corporate greed and global market failure. Cognizant of this and foreshadowing later criticism of the paradigm recurred in the form of the Olympic development, the writer is magnetized by César Pelli's pyramid:

> The seductive sky/water cemetery of Thatcherism, cloud-reflecting sepulchre towers: an evil that delights the eye (the eye in the triangle). An astonishingly obvious solicitation of the pyramid, a corrupt thirst for eternity. (Climb the true tower of St Anne's Church, and stand among Hawksmoor's crumbling Portland stone lanterns, pyramids set above catacomb arches, designed to be seen *through*, to keep vision alive; the river, all points of the compass – even the futile bluntness of Canary Wharf's phallic topping).[65]

Pyramids conflate. As Canada Tower was once the highest point in London, the tower of St Anne's, Limehouse, was once the highest clock in East London, visible from boats on the Thames.

Unlike Pelli's Canada Tower, Hawksmoor's time saturated construction, complete with pyramidal tomb to the west of the church, offers sustenance, 'keep[s] vision alive'. First, the visibility of time inscribed in the church-tower takes a loan on the 'sacred' authority invested in the buildings to perform an enclosure of the imagination through the disciplining of time. As the generative energy lines of Hawksmoor's churches mapped by Sinclair and Alan Moore suggest, and the silent word that relates to an elite, sectarian, Rosicrucian knowledge suggests, too, this discipline requires not just a clock but *a network of* clocks: not a singular force but an ecology of representation: an exemplification of embedded knowledge, of kybernétés (above) and heterotopia. Without this relation-driven network and temporal, contingent centric flow, the postmodern planners of Canary Wharf who 'have dabbled in geomancy [and] appeased the energy lines (while attempting to convert them) ... have achieved nothing beyond futile decoration'. Think surface, think Sinclair's Ackroyd. They offer nothing but another contemporary transformation from substance to form, locality to location. Conversely, as with Sinclair's use of Jones' relation to Hawksmoor, St Anne's makes things transparent and active, flowing: rather than heritage formation and the marketing of a near present within our financial grasp, we are informed that it is 'an image-generating time machine'.[66]

Occult Geometry and the Flow

New built environments that are 'blindly monolithic', consume the local environment, particularly infrastructure rolled out with the East London Olympic developments. The freshly laid grammar of public space, 'pavements' and 'bus-stops', according to Sinclair, 'aspire to an occult geometry of capital: Queensbridge Quarter, Dalston Square'; in these clinically transformed spaces 'everything is contained, separate, protected from flow and drift'.[67] As with his response to the Haggerston development, to be in the flow and to be in place – which has been emphasized by Sinclair's notion of (historical) orientation and navigating one's way through cultural formations – is not necessarily to be *authentic* but it is to lay a claim against capital. At least in these examples, to read capital as a mechanism that embodies certain norms and presupposes certain ways of valuing social and cultural integration, and affording limited modes of experience and movement most particularly. Moreover, Sinclair reads examples of the built environment itself as a unique cultural artefact in the sense that it 'both symbolically exposes the social relations that structure ways of life, and functions physically as a spatial system that reproduces them'.[68] The cynical occult geometry continues:

> No junk mail, please. No doorstep hawkers. No doorsteps. The big idea is to build in-station car-parks, to control "pedestrian permeability," so that clients of the transport system exit directly into a shopping mall. Where possible, a supermarket operator underwrites the whole development, erecting towers on site, so that Hackney becomes a suburb of Tesco, with streets permanently under cosmetic revision, replaced by 24-hour aisles. Light and weather you can control. Behaviour is monitored by a discreet surveillance technology.[69]

This sounds like a precocious critique of Stratford International, the extension to St Pancras' Eurostar infrastructure that has its platforms adjacent to the 180-acre Westfield retail development at the edge of the southern part of the Olympic park. Jardin du Luxembourg to London Fields, or Expo Paris Nord Villepinte to Stratford St Michaels (Marks and Spencer); different worlds (of commodities) in a more intimate space and time than that which these capitals have with their own suburbs.

But Sinclair's is also an aesthetic critique borrowing from the history of modern vernacular by Charles Jencks. The process metaphor of liberation of the modern can in the post-modern be 'appropriated by the vertically integrated and anonymous institutions of corporate and state capitalism';[70] while a celebration of technology becomes standardized for mass production manifesting in regular geometry and concrete hostility. It is also, as Jencks has regarded, forgetful and insensitive:

> It turns its back on the city and the past. It violates sensuous urban space with abstract forms and harsh angles; it consciously separates itself from urban society with an impermeable façade and a moat of windswept concrete or asphalt, and it eliminates historical reference, reducing building form to a rationalist, minimalist aesthetic which shows contemptuous disregard for function and place.[71]

To Sinclair's modernist mind, the centre of London's financial markets at Canary Wharf, must be referred to historically, as 'The Isle of Dogs', colloquial parlance signifying monarchical control of animals (Henry VIII kept his hounds here) while indicating postcolonial trade winds. The locus for corporate settlement restricts local action at the same time as it eliminates historical reference and contemporary reference to history; 'it repudiates graffiti ... we're trapped in an isthmus of signs, not language. A field of force deliberately set up to eliminate the freelancer, the walker, the visionary ... Systems of control based on necrophile geometry'.[72] Disregard for people, movement, intuition, pre-consciousness and pre-materialism; disrespect for place, function, signs for community, heritage, future.

Conclusions

The synthesis of communication and movement within a critique of hegemonic power that suppresses information flows, curtails human action (and movement) and thus fails to generate culture, reaches apotheosis when the writer under study is confronted with the 2012 Olympic development areas' security fences embellished with computer constructed projections of the finished site: here is '[the] future previewed, fixed, made inevitable'.[73] For Sinclair, ownership, access, usability of space, and its relation to capital, are axiomatic to understanding the dynamics of London (see Figure 2.2). Moreover, the history of limits, social segregation and exclusion and how these interrelate with the experience of capital and a specific part of the Capital is configured in contemporary literature as an act of delimitation and territorialization.[74] For geographers this is an exercise of the production of space, the social being defined somewhere between the political and the economic. It is clear that this is being rewritten in Sinclair as a privatized commons and enclosure of (historical) consciousness.

Figure 2.2: Road Closed Here, White Post Lane. © Nicholas Middleton

The capacity and the ability of the imagination to explore its environment over time are not pre-given; the built environment cannot alienate the very subjects they produce. However, this logic suggests only that there is nothing intrinsic about urban space that makes it alienating; true, but it is the rapid capital-oriented transformation of an environment that can prove unconducive to rich subjectivity and open identity and cultural formations. The fallout from this late capitalist power geometry that entails marketed and policed identity politics is hegemony, such that a 'mind' or 'subject' inscribed by 'one cultural milieu finds itself being written within another cultural environment involuntarily'.[75] Environments become heterotopic, then global, then homogenous; subjects become redundant centres of consciousness and cannot meaningfully connect with others unlike themselves, cannot regenerate through difference and relation. Preliminary inquiries suggest that market logic contradicts the project of building a common life. Here, in Sinclair's critical psychogeography, particular emphases on information flow and the resultant clusters of data that emerge and find representation are clarified through a critique of strategies and resources, which suggests that the new configurations of the public and private realms (which exploit heritage culture and regeneration policies and funding) are vulnerable - they can be opened up by an intriguing accumulative poetics which ties economic and social history to psychological and intellectual history. This poetic turn offers a new calligraphy of the city and a new history of its people and life-ways that rethinks geography as destiny.

Sinclair reminds us of the continuous play of history, culture, and capital; that culture is not autonomous but shot through with histories of interaction – as Foucault has said, it is not 'the undialectical, the immobile'.[76] Furthermore, this unfolding and ongoing sense of identity offers two further research questions. First, Sinclair's accumulative poeisis negotiates the opposition between synchronic and diachronic thought; is this due to an endeavour to assimilate experience within its interpretative structures (a method that enables it to learn from its past)?[77] Second, is this a form of radical situatedness where the enduring presence of the historical imagination is viewed through spatial analysis and dynamic fields of experience where 'space, ideology, and representation are joined in generative relations'?[78] The implicit, understated question herein, is perhaps the following: should these questions be left to historians, geographers, or aestheticians? Or to put this slightly better: how are humanities scholars best placed to think upon the habit of escape from meaning that is bound by institutional practices that have been devised to capture it?

3 HETEROTOPIA AND PLACELESSNESS IN BRIAN CHIKWAVA'S *HARARE NORTH*

Zoë Wicomb

Foucault's outline of heterotopias as counter-sites that relate to all other sites 'but in such a way as to suspect, neutralize, or invert the set of relations that they happen to designate, mirror or reflect' has, for all his extensive examples, been much criticized for its vagueness.[1] Moreover, critics point to the contradictions between the essay 'Of Other Spaces' and the brief reference to heterotopia in the Preface to *The Order of Things* where its focus is on textual space.[2] The arguments are pertinent to postcoloniality in that they centre around an understanding of Foucault's conceptualization of space as an instrument of resistance and social change: the doubleness and contradiction implicated in heterotopia has been linked with the postmodern valorization of alterity, and thus has inflected its meaning with radical 'openness'. Edward Soja, for instance, connecting the 'other space' of heteropology with Homi Bhabha's 'third space', insists that it is meant to detonate and deconstruct, that Foucault's geohistory 'explicitly focuses on the spatio-temporal interpretation of the power-knowledge relation'.[3] Hilda Heynen, on the other hand, is critical of heterotopia's doubleness 'which continues to resonate in between liberation and oppression.'[4] She cites Foucault's example of the heterotopic ship as a problematic site of imagination and adventure in its failure to recognize the ship's role in the history of slavery, racism and oppression. The arguments as to whether heterotopias subvert or support social systems are, however, deftly demolished by Johnson's detailed and nuanced reading of the concept as *not* being

> tied to a space that promotes any promise, any hope, or any primary form of resistance and liberation ... heterotopias are fundamentally disturbing places ... [they] draw us out of ourselves in peculiar ways; they display and inaugurate a difference, and challenge the space in which we may feel at home. These emplacements exist out of step and meddle with our sense of interiority. There is no pure form of heterotopia, but different combinations, each reverberating with all the others ... but their relationships clash and create further disturbing spatio-temporal units.[5]

It is such an understanding of heterotopia that informs my reading of Brian Chikwava's *Harare North*, a novel about migrants and asylum seekers in London that offers no hope of social change, but one that nevertheless trades in Foucauldian spatiality, in counter-sites where real sites are both represented and contested, being simultaneously 'outside of all places ... [and] absolutely different from all the sites that they reflect and speak about'.[6] Chikwava's narrative of impoverished, persecuted and displaced Zimbabweans exemplifies a globalized age of migration, unequal resources and of the restrictions imposed on refugees. Through textual analysis I will trace the ways in which heterotopia features in the depiction of a disturbing migrant identity and, in particular, consider Chikwava's use of intertextuality to portray London as a place of illusion.

Nomenclature and the act of naming in this novel is key to its hermeneutics. The first-person narrator remains nameless, and the name of the novel, *Harare North*, neatly embodies Foucault's third principle: 'heterotopia is capable of juxtaposing in a single real space several spaces, several sites that are in themselves incompatible'.[7] Thus Harare North is both the geographical site of contemporary London (as Zimbabweans call it), where a terrorized people have migrated, and a place in Zimbabwe. It is also, through intertextual reference to Sam Selvon's *The Lonely Londoners*, the heterochronic space of London during the first wave of black immigration in the 1950s. Thus the name exemplifies Foucault's identification of our epoch as one of both simultaneity and juxtaposition, and London/Harare North can be read as a heterotopic counter-site 'in which the real sites ... are simultaneously represented, contested and inverted'.[8] In *Harare North* nomenclature exemplifies the intersection of Harare and London, cities that are simultaneously occupied by the fictional Zimbabweans, in so far as their social lives in a temporary squat are essentially interactions with other Zimbabweans and are distinguished from their attempts at securing work in the metropolis. Marked by such placelessness, they are dismissed by hostile Londoners who 'think you is in the wrong place but don't tell you straight and square'.[9]

Foucault's very examples of heterotopia outlined in 'Of Other Spaces' such as the mirror, museum and cemetery occur in this text, but the novel at the same time appeals to textual space as referenced in *The Order of Things*, thus cutting across the commonly perceived contradictions. However, if *Harare North* lends itself to a 'heterotopic reading' it is manifestly the case that heteropology cannot be *applied* to a novel or seen as a hermeneutic tool. Instead, this novel highlights as Johnson points out, the fact that 'Foucault was profoundly influenced by his reading of the spatial dynamic of literature' and in particular the 'space of literature' and placelessness as discussed by Blanchot.[10] It is precisely Chikwava's carving out of such a space and his shaping of its relationship to the imagination that drives the novel. Johnson, however, ultimately cannot resist reading heterotopia as an appeal to resistance and transgression. He concludes that in some ways the concept does provide 'an escape route from power', that it is liberatory in so far as it illuminates 'a passage for our imagination'.[11] But for Chikwava's migrants, caught between repressive powers, no escape route is on offer; theirs is indeed an illusory flight from oppression, and the indeterminacy of the novel with its ambiguous ending illuminates rather the ontological dangers for a protagonist stripped of everything but a hectic imagination.

The central theme of the novel is betrayal, which occurs in a number of intersecting heterotopic spaces. Primarily, the reader is alerted to the betrayal of the liberatory postcolonial ideal in Zimbabwe so that the refugees in London have no hope of return to the homeland or of recovering that ideal. Resistance is a thing of the past, a force by which a once revolutionary Mugabe set out to liberate the country from colonialism, and which is still embraced by deluded Zimbabweans in whose interest it is to not recognize the betrayal. One such is the unnamed central character. He has led a 'rubbish life' in Harare before being recruited by ZANU-PF party who 'give you chance to change your life and put big purpose in your life'.[12] Thus he refuses to acknowledge Mugabe's failure, persisting in the idea that opposition to the state is treacherous, and the diegesis of the narrative tracks his slow understanding of political betrayal. As a member of the ZANU youth movement, known as Green Bombers, or 'jackal boys', he has fallen foul of the police; he comes to London for the single purpose of earning $5000 as a bribe with which to return to Zimbabwe, believing that his misdemeanours have in any case been the legitimate actions of one who acts against traitors to Mugabe's regime.

At the end of 'Of Other Spaces', Foucault ponders the well-arranged spaces of colonies functioning as heterotopias of compensation for the messy, ill-constructed and jumbled metropolitan spaces of the West. Chikwava views the space of London from the other end of the colonial telescope: the narrator both challenges and confirms Foucault's distinction between illusory and compensatory heterotopias. If contemporary London exemplifies the description of the world as one of both simultaneity and juxtaposition, 'the epoch of the near and far, of the side-by-side, of the dispersed',[13] Chikwava's post-colony is also postlapsarian. Zimbabwe is anything but a well-arranged space; indeed it has physically shifted to London where the main employment available to migrants is what they call BBC work (British Bum Cleaners), or home help for the elderly.

The opening scene of the novel is set in Gatwick airport, designated site for arrivals and departures, or the management of movement, but in the age of global mobility also a heterotopic place of detention. Chikwava's anti-hero, who knows that he is expected to denigrate the Zimbabwean government, falsely claims to be a persecuted member of the opposition and thus in need of asylum. Detained all the same for eight days, the protagonist comments on the irrationality: 'Whatever they reasons for detaining me, them immigration people let me go after eight days'.[14] What sets this novel apart from other narratives about asylum-seekers, is the history of the narrator. This anti-hero is not victimized by the regime. With his anti-democratic past he has no claim to the sympathies of the metropolitan reader, and the humour of the text springs largely from his skewed understanding of politics. He is uneducated, and as a product of Mugabe's misrule believes in the duty 'to smoke them enemies of the state out they corrugated-iron hovels and scatter them across the earth'.[15] Such a biblical view is an ironic take on globalization. As the story unfolds, we see that there is little to choose between the dispossessed, hungry, betrayed refugee opposed to Mugabe or those pressed into support for him, and that irrational policing at the airport will in any case not make just distinctions.

Having waited another two days after his release before Sekai, his cousin's wife, collects him from the airport, the narrator is struck by his relatives' behaviour as 'lapsed Africans'. Their deracination is marked by newly acquired metropolitan manners such as caring for a dog, eating ready-made meals and their commitment to DIY home improvements. Striving to be absorbed into London, they are reluctant to put up with him, thus he moves out to a squat in Brixton where his childhood friend, Shingi, lives with a group of other Zimbabweans. The narrator holds reprehensible patriarchal views, such as believing that his cousin could save his marriage by keeping his wife occupied with a baby, or that lesbians lack real men and could be cured in one night by someone like himself. His ignorance about AIDS persuades him that being certified HIV *negative* cannot be good news. Nevertheless, he engages the reader's empathy, primarily through his wayward intelligence, his linguistic skills, but also through the geohistory conjured by the text in which, as Soja has it, 'the ontological trialectic of spatiality, sociality and historicality' is explored.[16] For all the robustness of the characters in this novel, migrant subjectivities in an unreal city are fragile, and the novel ends with dissolution of the self, where the unreliable narrator would seem to merge with the character of his childhood friend and compatriot, Shingi.

In this first person account, non-standard English allows the narrator to refer to himself simultaneously in both the subject and object form. In an emphatic register he typically states, 'me I find no reason to continue' or 'me I don't see what the whole noise is about',[17] thus signalling a doubleness that develops in the course of the novel. The ontological crisis into which the cocky anti-hero slides is bound up with the nominal effect that allows his identity finally to slip into the doubleness of being both himself and his friend Shingi. And the unreliable narration does not make it clear at what stage his identity merges with Shingi's. Heterotopias, as Foucault warns, 'are disturbing, probably because they secretly undermine language, because they make it impossible to name this *and* that, because they shatter or tangle common names'.[18] Significantly, the narrator does not name/rename himself; it is the reader who through his confusing discourse and tangled identities finally attaches the name of Shingi to him. The suspension of the narrator's name, as Derrida also states of the unnameable in his discussion of Babel, is a play that makes possible nominal effects ... the chain of substitution of names in which, for example, the nominal effect *différa*nce is itself *enmeshed*, carried off, reinscribed, just as a false entry or a false exit is still part of the game, a function of the system.[19]

Doubleness then is also inscribed in other names of characters that readers are invited to substitute: The narrator/Shingi/ for Moses, the central character in *The Lonely Londoners*; 'Original Native' for Shingi; Tsitsi, the Zimbabwean girl in the squat, for the narrator's Mother; and more playfully, the narrating I for Sherlock Holmes, represented in the novel via the heterotopic space of the Holmes museum. The dissolution of the self, or madness, into which the beleaguered narrator descends, follows Shingi's earlier drug-induced craziness and a knife attack that leaves him physically and mentally impaired. And the dissolution of the narrative discourse itself echoes Foucault's description of disturbing heterotopias that 'destroy "syntax" in advance, and not only the syntax with which we construct sentences but also that less apparent syntax which causes words and things (next to and also opposite one another) to "hold together"'.[20] By the end of the novel it is precisely such cohesion that fails; names are indeed tangled as the unnamed narrator 'recognizes' Shingi in his reflection, and the names of his cousin and uncle become tangled with those of Shingi's relatives. The syntax of the narrative discourse is, of course, necessarily rule-bound; however, it is the 'less apparent syntax', the order of things, the distinction between real and imaginary that is shattered, as in the case of the Museum at 221b Baker Street, visited by the narrator at the end of the novel. This house where Sherlock Holmes and Dr Watson lived is, as its promotional literature claims, faithfully maintained for posterity as it was kept in Victorian times, but whilst the space is real in its reconstruction of Holmes' residence, it is of course the case that Holmes and his home are fictional, hence heterotopic. And the conjoined lives of Holmes and Watson, as well as Watson's role as chronicler of Holmes, also draw attention to the I/Shingi conjunction that emerges at the end of the novel. Moreover, Sherlock Holmes adopts disguises in the course of his investigations; he is often portrayed as patriot acting in the interest of national security, mirroring the narrator's misguided belief in Mugabe; and like the I/Shingi he is a user of addictive drugs. When the narrator is told in the museum that Holmes is a fictional character, he leaves in disgust, seeking out instead his own image in a mirror.

Nomenclature and doubleness are bound up with the theme of betrayal. The Commander of the Green Bombers, Comrade Mhiripiri, a father figure who in Zimbabwe claimed to negotiate the bribe with the police and to whom the narrator writes for help, has in fact migrated to London. He is disguised as the self-seeking vagrant who loiters at the chestnut tree in Brixton where various refugees congregate. When the narrator confronts Mhiripiri, the reader is momentarily unsure as to whether the man, known in Brixton as the Master of the Fox Hounds (MFH) whom he had encountered before, is indeed the duplicitous

Commander, or whether the narrator is deluded. In other words, the moment of anagnorisis is itself ambiguous. However, it is the popular name of the Green Bomber unit, Boys of the Jackal Breed, that confirms the MFH's identity, and the earlier, puzzling comment on the MFH when the narrator first meets him at the tree – 'he have change'[21] – confirms the moment of recognition as delayed or dissipated anagnorisis, concealed from the self as the anti-hero is reluctant to face the truth about the fraudster who had planned to pocket the bribe. Recognition of the MFH is finally achieved through addressing him, but instead of embracing the truth about Zimbabwean politics, the narrator clings to his regressive views.

The rhetorical devices employed in the text (for example, in relation to the narrator's absent mother as cipher), carve out yet another kind of heterotopic space, one which Foucault glosses as follows: 'words have their locus, not in time, but in a space in which they are able to find their original site, change their positions, turn back upon themselves, and slowly unfold a whole developing curve: a tropological space'.[22] The narrator's political and moral blindness is poignantly demonstrated in his reception of the news about his mother's village; indeed, narrative time and the diegesis are marked by various references to the regime's mining of emeralds and the destruction of the village including its cemetery where his mother is buried, accounts that he dismisses as propaganda against the state. Mother (capitalized throughout) is first introduced via the suitcase with which he arrives in London and which marks him as Other. His metropolitan relative is embarrassed by its shabbiness, but for the narrator the suitcase is a repository of history and memory, less of an object than space, and variously deployed in the narrative, it becomes a site of tropological space. On his arrival in London he describes the suitcase as 'one of them old-style cardboard suitcases that Mother have use before I was born and have carry roosters in the past, but it's my suitcase. It still have smell of Mother inside'.[23] Thus the suitcase, which accompanies him in the crossing over from Harare to London, serves as metonymic link to his mother, representing both continuity with the past as well as rupture from that time and that place. The suitcase accrues symbolic value: in the squat in Brixton the narrator keeps it locked and guards it closely. Often it serves as contemplative space: he sits on it looking out of the window, whilst writing his diary and letters; it is the place where his and Shingi's money is kept – precious commodity that will achieve his ambition of leaving London and returning to Zimbabwe.

The dominant trope in the novel is metonymy, displaced through analogy, adjacency and various associations. At the end of the novel, the deranged narrator walks the streets of London with the suitcase on his head, representing simultaneously an earlier era of a traditional mode of carrying. Thus it exemplifies Foucault's claim of heterotopias being linked to heterochronies, the break in traditional time that is also manifested in cemeteries where a quasi-eternity has its beginnings in loss of life. Existing then in both a colonial past and a postcolonial present, the suitcase, bearing the deceased Mother's smell, replaces her 'place of rest'. At the same time, in olfactory terms, it is the site of her resurrection. Heterochronically marked, the suitcase functions like Foucault's example of the boat, a place without a place that goes 'from port to port ... goes as far as the colonies in search of the most precious treasures'.[24] But in this case, viewed from the postcolonial subject's contraposition, the narrator's quest for money that will make possible a return to the homeland. The 'smell of Mother', which pervades the entire text, thus bears the pathos of an absence made present, and through strategic use of nomenclature occupies the suitcase in both Harare and London.

In relation to the skewed morality that confirms the unreliability of the narrator, Mother performs an important tropological function through which, ultimately, the ethical is inscribed. Structured absence in the text, she is abstracted into motherhood, and thus as concept metaphor embraces also Shingi's biological mother who had been raped during the war, and Shingi's foster mother who at the end of the novel, as the protagonist becomes fused with Shingi, also becomes fused with Mother. The narrator's mother who first appears in the text through metonymic displacement via the suitcase that carries her smell, is both there and not there. Her death is after all rendered ambiguous by the fact that her son has failed to perform the traditional *umbuyiso* ceremony, which should have laid her spirit to rest. Instead, it is wandering in the wilderness – which here residing in the suitcase proves to be yet another name for London.

Alienated early in the novel by his relatives, Paul and Sekai, the narrator takes comfort from thinking about his mother who since her funeral '*have knit sheself back into life*' (14). These thoughts or fantasies, presented in italics on pp.14–16, resurrect Mother after the socially anxious Sekai admonishes him for cleaning his shoes on the doorstep as if he were in a township. Cohesion between the real and the imagined scene is achieved through substitution of shoes – the fantasy of being home starts with his mother's old shoes on the doorstep, '*wet and red with mud*'.[25] The fantasy is marked with other such realist operators as conversations with neighbours; the mother's housework, which includes dusting a display cabinet with a photograph of her son; and the red hens that she knits, '*the same colour as jersey that I am wearing in photograph*',[26] thus linking him to a distant past. But she also boasts to her friends about her son in *Harare North* who in another photograph is feeding pigeons in London – in other words, the fantasy itself deals in a contradictory temporality. It repeats the contradictory spaces of the novel's title: in London, Mother is revealed in her Zimbabwean house where London is represented in an iconic photograph. A reference to Selvon's *The Lonely Londoners* (already signalled in the repetition and difference of the opening scene of arrival in London) is slipped in when one of the neighbours complains that the protagonist feeds pigeons in Harare North whilst people are starving back in Harare.[27] The spatio-temporal reality is thus further disturbed through intertextuality.

The conversation between Mother and her friends collapses into cacophony and chaos, a disturbance that ranges to spatial shifts between inside and outside, achieved through sound: the 'other sounds' of crows, winds tearing through the garden, windows banging, and the clanking of knitting needles. It is within this disturbance that a conversation between mother and son, who in the process regresses to infancy, becomes possible, a scene of painful interiority, of infinite tenderness, longing and loss, where the naming and repetition, the call and response in the opening sequence places it in a past of traditional discourse that ends with its irrecoverability.

Mother, I hold she tight in me arms.

'My child', she cry.

'Mother'.

'My child.'

'Yes, Mother?"

'Mwanangu.'

'Amai.'

Mother. She wrap me up in she arms and hold tight. My small feet lock together, them small toes coil. I'm back in Mother's arms.

'Did you fall, my child?'

I suck thumb and nod. Mother hold me to she bosom and rock me gentle. Then some funny long breast roll out down and swing past my face like pendulum ... Outside, things is now quiet. Inside, breast is cold; the milk dry up long ago.[28]

The oneiric richness of this scene also alerts us to the ambiguity of the child's fall – the moral lapse in the narrator's embrace of Mugabe's misrule. And the large symbolic breast serves to mobilize the script of revered Mother Africa, one that Chikwava deploys in his gentle mockery of tradition, and in particular to show the violence against women that goes unquestioned in the name of pre-colonial tradition.

The capacious concept of Mother Africa (woman as guardian, nurturer, repository of unconditional love, largesse) is later deconstructed through a series of situations and events: Tsitsi, the girl in the squat is a mother, but also is a mere child. She has been abused and is still exploited by another housemate, the duplicitous Aleck, just as Shingi's mother had been raped by a pre-independence freedom fighter, and to add insult to injury, was then 'taken away ... to punish Shingi for the sins of his father'.[29] The protagonist's own mother's grave is desecrated as Mugabe's soldiers raze the village in order to mine emeralds. In other words, the very reverence for motherhood, celebrated in the name of tradition is in the postcolony shown to be cant. The fantasy reveals more than the narrator's regression to infancy: it suggests a wider national breakdown where the shift from nurturing to retribution is registered through the pendulous, dry breast.

The naïve narrator, unable to interpret the vision, remains enthralled by the concept of motherhood. He is fascinated by Tsitsi caring for her baby 'like real mother'[30] and defends her against Aleck on the grounds that 'you don't do that to mother'.[31] But Tsitsi herself betrays the ideal of motherhood by renting out her baby to others for claiming state benefits, and ultimately as his money runs out, the narrator in turn betrays Tsitsi by pressurizing her into leaving the squat. This betrayal is prefigured in a dream where his own mother merges with Tsitsi: as a member of a band of hungry Green Bombers who terrorize the villagers, he encounters an old woman who resists, and who turns out to be his mother: 'she say we should give back all them chickens and learn to give instead of being useless ... We is leaving empty-handed and I look back at Mother but she face is now Tsitsi's'.[32] When he finally breaks down and leaves the squat with the suitcase on his head the smell of Mother has seeped from the suitcase to fill the entire house.

The protagonist's distorted understanding of Zimbabwean politics is poignant in the face of the current eviction of villagers and destruction of their cemetery by the Green Bombers commander who also owns the mining company. Foucault's example of the cemetery as heterotopia, an example of breaking with traditional time, or the way in which 'a society, as its history unfolds, can make an existing heterotopia function in a very different fashion'[33] is given an ironic twist as the mother's grave is desecrated, and the Zimbabwean cemetery collapses into emerald mine. This crisscrossing of space and time, of burial and exhumation, is reminiscent of Blanchot's discussion of the strangeness of a cadaver:

> The deceased rests heavily in the spot as if upon the only basis that is left him. To be precise, this basis lacks, the place is missing, the corpse is not in its place ... It isn't here and yet it is nowhere else. Nowhere? But then nowhere is here ... The cadaverous presence establishes a relation between here and nowhere ... doubled by himself ... joined to his solemn impersonality by resemblance and by the image ... he resembles himself. The cadaver is its own image.[34]

Blanchot's cadaver is pertinent to the representation of placelessness in Chikwava's novel. We learn that the *umbuyiso* ceremony, which brings the spirit of the deceased back home, has still not been performed for Mother. Her grave, desecrated by bulldozers, means that Mother literally has no resting place. Doomed to permanent wandering in the wilderness, she now 'occupies' placelessness, which by the end of the text is also embraced by the narrator who in his doubleness seeks out and becomes his own image.

It is the narrator's reluctant recognition of the rottenness of the state that heralds the dissolution of the self, so that his identity slides into that of his childhood friend, Shingi. Shingi, in turn, although he has been granted asylum, is betrayed by the promise of freedom in Britain where he falls prey to drug abuse and, savagely attacked by vagrants, is left physically and mentally disabled. But the narrator's account of Shingi is unreliable. The italicized text on p. 195 about his comatose friend brought home in a wheelchair is not to be trusted in the light of his later concern about not having been to visit him in hospital. Instead he engages in comic correspondence with members of Shingi's family who beg for money, deftly discriminating between those who genuinely need money and those who are greedy. Deranged by further reports of oppression in Zimbabwe, he decides to leave the squat and to wash his hands of the problem of Shingi, a decision that turns out to be one of abandoning the self, since it is the image of Shingi that faces him in the mirror. The novel ends with the narrator – who earlier claims to be 'as healthy as a Harare North dog' – descending into bizarre behaviour and confusing language. He cuts an eccentric figure with the suitcase on his head, wandering the streets of London, and keeping to the white line in the middle of the road where he feels safer. This line divides the flow of traffic; in terms of the traveller it is neither here nor there; being inbetween it signifies neither coming nor going; and it is less about the narrator's liminality than about positioning himself in placelessness, or a heterotopic nowhere. It is here that he finally recognizes himself as pariah; he feels like *umgodoyi*, the homeless, persecuted dog. He only puts the suitcase down when he realizes that the lock has come undone, that his possessions have been scattered: 'Nothing is left inside suitcase except the smell of mother ... it's full of nothing.'[35]

The portrayal of a migrants' London relies heavily on the use of intertextuality. Chikwava's self-conscious referencing of Selvon's *The Lonely Londoners* is instrumental in representing the city in heterochronic terms. Intertextuality then, as deployed in *Harare North* departs from the 'writing back' of canonical postcoloniality, the ironic re-presentation of the native in colonial texts. Eschewing the conventional function of re-presenting the colonial vision and restoring subjectivity, Chikwava references Selvon as palimpsest through which to offer a more complex representation of migrant history and space. If intertextuality is, as theorists claim, a condition of all writing, its significance for the postcolonial lies in its transformative effect. Repetition of Selvon shows that the plight of the migrant has remained the same, but repetition-with-a-difference also shows how that plight has been radically transformed. Heidi Sohn in her analysis of Foucault's heterotopia identifies 'transformation of meaning as the common denominator to access his description,'[36] a notion that underlines the spatiotemporal disruptions that citations of Selvon's *The Lonely Londoners* bring to Chikwava's novel. Whilst referencing the same place, the very name of the novel, *Harare North*, i.e. London, constitutes a shift from naming people to naming place, and not only does it omit both the affective qualifier and the definite article, but in the substitution destabilizes place by keeping its referent secret.

As an update of migrant experience, *Harare North* appeals to literary history, making a heterochronic link with the first wave of immigration to Britain in the 1950s, and by introducing dialogue between the two texts, exemplifies the spatio-temporal inter-connectedness, or as Foucault has it, our experience of the world as a network 'that connects points and intersects with its own skein'.³⁷ Thus we see in the text, 'juxtaposed in a single real space several spaces, several sites that are in themselves incompatible'.³⁸ It is 2009 and the fraught, humiliating business of entering Britain is a far cry from the relaxed jolliness of the boat train by which Selvon's immigrants arrive at Euston station in the 1950s. If the meaningful social lives of West Indian immigrants are staged around a gas fire in Moses' room where they congregate on Sundays, this safe, congenial space is transformed. In 2009 Chikwava replaces it with a chestnut tree in Brixton – 'all them funny types is gathered there, them the crazy ones and them the ex-pig keepers who have flee they crazy countrymen in hot climates'.³⁹ The narrator distances himself from those who 'start to trickle to the tree with they dogs, ready to start to put out the burning truths of they lives with buckets of brew and all'.⁴⁰ But the contemporary migrants are heterochronically linked to the West Indian immigrants of the 1950s through the chestnut tree, the space where they congregate to tell their stories which is simultaneously the intertextual site of Moses' room.

Chikwava's literary precursor is nowhere more evident than at the end of the novel when his narrator ponders on what to do about Shingi. He goes to the river where, sitting under the bridge, he resolves to abandon his friend. Later we discover that it is from Waterloo in the early morning that he catches the tube to Brixton, echoing and also reversing the journey made by Moses who on a winter's evening goes by bus to Waterloo to meet newly arrived immigrant, Galahad, at the boat train. (For the unnamed narrator, it turns out to be a meeting with the self in the guise of Shingi.) But the scene also echoes the ending of *The Lonely Londoners* where '[t]he old Moses, standing on the banks of the Thames' contemplates his life in London, 'the swaying movement that leaving you standing in the same spot. As if a forlorn shadow of doom fall on all spades in the country'.⁴¹ Decades later it is the same shadow that falls on Chikwava's narrator whose experience of London as contradictory place is confirmed. The biblical name of Moses indicates that Selvon's character will not reach the promised land – he is in any case ambivalent about returning to the West Indies – but unlike Chikwava's that novel ends on a warm, hopeful note:

> it had a greatness and a vastness in the way he was feeling tonight ... It was a summer night: laughter fell softly: it was the sort of night that if you wasn't making love to a woman you feel you was the only person in the world like that.⁴²

It is also the night that Moses thinks of becoming a writer.

It is both continuity and reversal that Chikawa achieves in his use of *The Lonely Londoners* as intertext. The shadow of doom suggests that our narrator who, unlike Moses, passionately wishes to return to Harare is unlikely to do so. However, instead of the sanctuary that is Moses' room, the Zimbabweans' squat, both a refuge and a place of conflict where Aleck betrays his compatriots, is now empty, and the narrator descends into madness, or at least into his double, Shingi. Chikwava also exploits the nominal effect of Selvon's Moses. If we are invited throughout to link Moses with the unnamed narrator, the biblical identity of Moses as a stutterer is repeated in Shingi's speech impediment, thus the name is exploited to indicate the doubleness of the Shingi/narrator identity.

Reference to Moses' musings on writing – 'One day you sweating in the factory and the next day all the newspapers have your name and photo, saying how you are a new literary giant'[43] – is repeated in *Harare North*, and not only as a self-reflexive gesture. The nameless narrator writes a diary, which could be seen as something akin to the text we are reading. But he also dreams of meeting a ghost writer who will transform his diary into a bestseller so that he can return to Zimbabwe and perform a lavish *umbuyiso* ceremony for his mother. He starts assuming Shingi's identity through writing, reading and replying to Shingi's letters from relatives, and it is in these letters and the money he sends to the deserving ones that we see Moses' compassion replicated. The ethical imperative for the reader is to accept the narrator, for all his scheming and reprehensible affiliation, as himself a victim of poverty, ignorance and an exploitative system, whose thoughtfulness, humour and generosity go some way towards eclipsing his political blindness.

It is towards the end of the novel, where a paragraph on the first page is repeated, that we discover its invaginated structure. In the Prologue, Brixton is described as follows:

> When I climb out of Brixton tube station that morning, there is white, ice-cold sun hanging in the sky like frozen pizza base. Beyond the station entrance, some chilly wind is blowing piece of Mars bar wrapper diagonal over pedestrian crossing. And the traffic lights – they is red like ketchup.[44]

This desolate scene relies on references to food, the precious commodity over which the migrants squabble, the unfulfilled promise of plenitude in the metropolis symbolized in the inedible, inaccessible pizza base, devoid of its topping. Later in the text the protagonist, consulting the internet, finds that immigrants' 'contribution to this country is equal to one Mars bar in every citizen's pocket every year' (24), thus here at the beginning of the novel the empty Mars wrapper indicates his uselessness in the host culture. After the lengthy analepsis that constitutes most of the novel, we find towards the end the following variation:

> When I climb out of Brixton tube station, some pale icy sun hang in the sky like frozen pizza base. In them, these mental streets, bitter cold wind is blowing. And the traffic lights – they is red like ketchup.[45]

Now back in base time, the place is as desolate as ever, but the Mars bar wrapper has disappeared; it is a currency that no longer interests the narrator, whose mental state has been transferred to the streets. Our attention is drawn to his madness. The red light of prohibition remains; London is a place of restrictions; in this placelessness there is nowhere to go. This scene, which is also both beginning of the story and close to the end of the narrative discourse, starts on a day when he rises early to make his decision. What was unclear at the beginning, the decision to discontinue his friendship with Shingi – another betrayal – is shown at the end of the novel to be realized, paradoxically, through 'becoming' Shingi. In other words, the narrator's empathy renders him incapable of betraying his friend. The concept of betrayal is itself re-semanticized, its transitivity scrambled, since the notion of an other implicated in the process of betrayal is radically transformed: the other is at the same time the self, the old Green Bomber self who would not face the truth, and who has from the start been betrayed by those in power.

We should not have been so surprised by the merging of identities, since for all his adherence to ZANU-PF, there has throughout the text been indicators of loyalty and love for Shingi that border on slippage into the latter's identity. Early in the novel, as the narrator sits in the toilet of Paul and Sekai's house (another heterotopic space of taboo that at the same time functions as sanctuary) thinking about his old friend Shingi, he declares, significantly using the second person: 'you know each other so well that sometimes you is not sure if your memories belong to him or vice-versa; things can get mixed up and time become one tangled heap and you no longer know whose story belong to who'.[46] In reminiscing about their childhood and the food game they played, he claims to have always won, and concludes: 'I possess him. I still possess him',[47] suggesting their doubleness whilst ironically prefiguring the fact that it is Shingi who will take possession of the narrator. At Tim's Fish Bar where he finds work, he uses Shingi's passport and National Insurance number. This is both a necessary ploy by hungry immigrants, as well as an illustration that in the host culture all blacks look the same.

The narrator's assumption of Shingi's actual identity starts with a mirror, a device cited by Foucault as an example of heterotopia's disruption of space. Foucault, in his discussion of Manet's *Un Bar aux Folies-Bergére* comments on the distortion between what is represented in the mirror and that which ought to be reflected,[48] an exploration of doubleness that is developed in 'Of Other Spaces' where the mirror exemplifies a third space. According to Boyer, Foucault's mirror opens up questions of how the mind conjures up imaginary worlds; it is 'a space of comparison between the virtual image in the mirror and the image of the self, comparison between an image of utopia and dystopia, the past and the present'.[49] In Chikwava's novel the mirror is explicitly linked to textuality and orthography. After Shingi has been assaulted and hospitalized, the narrator goes into a shop where the following notice is written on a mirror: 'This mirror compresses your image and makes you look short, squat and wide. We suggest you go to the basement where there's a better mirror that will make you look nice'.[50] The notice is reproduced in the text in bold script, foregrounding the placelessness where the narrator with a suitcase on his head is positioned, between the dystopic and utopic images. He leaves the shop, but slippage between the two identities is soon achieved through a puddle of water that serves as mirror in which the narrator sees

> Shingi looking straight back ... My stump finger now feel cold and sore from carrying suitcase. I shake my head and Shingi shake his head until I start to feel dizzy. Why he want to shake me out of his head like so, me I don't know.[51]

In Foucault's terms the common names have indeed been tangled, with Shingi's stump finger now seamlessly attached to the narrator. The narrator has already started assuming Shingi's identity earlier through writing letters to his family; he has cracked his friend's email password, a predictable 'Original Native'. In the early stages of remembering their childhood, he recalled Shingi back in Zimbabwe licking ice lollies, resisting 'everything the weather throw at him' so that he is admired by his friends as 'the Original Native'.[52] Shingi's cousin and uncle with whom he has been in contact via letters and telephone calls now turn into the narrator's relatives, so that the reader participates in the problem of identifying characters. The namelessness of the protagonist is reinforced by a pronominal shift. Significantly, the first person narrative agent, characteristically both subject and object as in 'me I don't know' now shifts to second person:

> You start to hear in tongue; it feel like Shingi is on his way back to life ... You can tell, you know it; Shingi is now coming back ... You stand there in them mental backstreets and one big battle rage even if you have no more ginger for it.[53]

The inclusive pronoun, 'you', an ambiguous narrative agent, confirms grief and empathy to be instrumental in the slippage of identities. The nameless subjective 'I' that gives way to 'you' signals not only a merging of identities but a death of the self that is at the same time a resurrection of the physically and mentally damaged Shingi.

Intertextuality is significant at this stage in layering meaning and in reinforcing similarities whilst simultaneously asserting difference from the early experience of migration: London is and is not the same place. Shingi, the Original Native, is textually linked to Galahad in *The Lonely Londoners* who, on arriving in London in midwinter, is scantily dressed and throughout remains oblivious to the cold. He is, however, at the same time associated with Selvon's Moses through his stuttering. Selvon too, in his non-standard English focalized primarily through Moses, strategically shifts to second-person narration in a lyrical, unpunctuated stream-of-consciousness passage about summer in London – ('Oh what a time it is when summer come to the city').[54] Through the inclusive 'you' Moses' subjectivity is shown to be bound up with the rest of the immigrant community; conversely, Chikwava's usage of the second person indicates dissolution. An earlier second person usage in *Harare North* is anything but inclusive; rather, it registers distance and otherness as the narrator in free indirect thought regards the hostility of native Londoners: 'You see me stepping down them pavements ... You give the talking eye that demand your Mars bar? You lick ice cream, I bite mine'.[55] At this stage he has had his first pay packet and shares his money with Shingi, thus prefiguring their doubleness also through reference to ice-cream and Galahad's tolerance of the cold.

Through persistent reference to Selvon, Chikwava insists on a heterochronic coevalness that speaks of the permanent lot of the migrant. This he does through repeating lexical items like the non-standard use of 'coasting', a word that broadly describes the distinctive postcolonial wandering through the city; through substitutions like 'laugh kak kak kak' for Selvon's 'laugh kiff-kiff'; 'yari yari yari' or 'spin some jazz number' for 'big ballad'; and the substitution of the refrain, 'Me, I am no civilian' for Moses' characteristic utterance, 'Take it easy'. There are also events, both analogous and contrastive, like the arrival in London or interactions with native Londoners, as well as analogous characters like MaiMusindo who echoes Selvon's Tanty and at the same time radically departs from her. These assert simultaneously the sameness and difference between contemporary London and the city of the fifties; it is, as Foucault maintains, the relational aspect of heterotopia that signifies. Thus Harare North is a fundamentally disturbing place that draws our narrator out of himself, in the manner of his own idiosyncratic rhetoric, so that the space of the self is taken over by Shingi.

The textuality of heterotopia that Foucault outlines in *The Order of Things* is here at play. As in Blanchot's 'Space of Literature' which implies 'withdrawal of what is ordinarily meant by "place"' and which also suggests the site of this withdrawal ... like the place where someone dies: a nowhere ... which is *here*', Chikwava's London comes to represent placelessness.[56] It is within such spatiality and in the imagination of the narrator that a tropological space is forged. For instance, he describes Shingi's squat as a house with sad eyes and a bay window that sticks out like a nose: 'When I look at the nose, the eyes, the black parapet wall – this is Shingi straight and square. But you don't tell anyone that they head look like a house'.[57] Our narrator's ears are attuned to what Foucault calls 'the murmur of analogies rising from things'[58] so that the simile multiplies in metonymic currency: he later tells Shingi's alarmed relatives that he could meet them inside Shingi's head, meaning at Shingi's house, but at the same time disturbing the facticity of the scene; and when the electricity meter runs out of credit 'there is darkness inside Shingi's head'.[59] The same device is used in the comic representation of immigrants' contribution to the country as the equivalent of one Mars bar in every citizen's pocket every year. This graphic representation of an ameliorative Mars bar is converted into a flexible metonymy where the narrator sees in natives' hostile eyes their demand for their Mars bars. The narrator's rhetorical skill then not only wins him sympathy from the reader; it contributes to the representation of heterotopia, where the incompatible sites of Harare, contemporary London and London of the fifties are juxtaposed in a single real space.

If *Harare North* ends with dissolution rather than resolution, and the lack of self-knowledge on the part of the narrator is not clearly restored, the figure who walks the streets of Harare North with a suitcase on his head, carefully keeping to the white line in the middle, is one who regardless of his politics and regardless of our expectations of redemption, cannot fail to engage our sympathies. We have seen how intertextuality contributes to the space of literature, which is glossed as follows by Blanchot's translator:

> it is its very own displacement ... it shelters nothing within it; it is also called *le vide*, the void. Sometimes it is associated with the anonymity of big cities, sometimes with the gap left by the absence of the gods, but sometimes too with what Rilke calls the 'world's inner space'.[60]

The *nowhere* or *placelessness* of the white line in the road that our narrator with a suitcase on his head treads is emblem for such a space. And his careful movement along this line signals the time for the novel to end. The nameless narrator, invaded by his friend Shingi, can now be named. If through his central character Chikwava has celebrated the rhetorical dimension of words, Foucault reminds us that rhetoric is bound up with the space that surrounds the Name, with 'the figures through which discourse passes as a deterrent to the name, which then arrives at the last moment to fulfil and abolish them. The name is the *end* of discourse'.[61]

4 'IT'S A FREEDOM THING': HETEROTOPIAS AND GYPSY TRAVELLERS' SPATIALITY

Mariangela Palladino

> Space is fundamental in any form of communal life; space is fundamental in any exercise of power.[1]

Cultures are organized spatially, they must 'remain in the place they belong, and only there'.[2] This sedentarist metaphysics, as Kabachnik points out, is the 'hegemonic norm ... it is seen as natural and taken for granted'.[3] Sedentarization entails exercising authority through and across space, it is a dominant, ethnocentric instrument of power. As a spatialization of social order, sedentarization belongs to the realm of what Lefebvre calls '"dominated" space', 'a site of hegemonic forces'.[4] As Lefebvre puts it: 'space has become for the state a political instrument of primary importance ... It is thus an administratively controlled and even policed space'.[5]

Sedentary society is preoccupied and 'burdened' – to borrow a post-colonial term – by the settlement, regulation, ordering and containment of its others, its nomads, by the ever romanticized, and demonized, wandering Gypsies. 'It is common knowledge', Deleuze and Guattari say, 'that nomads fare miserably under our kinds of regime: we will go to any lengths in order to settle them'.[6] The legal rhetoric concerning nomads ranges from hostility to paternalism, yet it remains consistent in aiming at sedentarization.[7] Indeed, from the sedentarist perspective, the spatialization of Gypsy Travellers[8] escapes the settled logic, it entails a rupture from the dominant, metropolitan imaginary.

By focusing on the representations of Gyspy Travellers' sites and dwellings, this chapter explores spatialization in the context of mobility; it will analyse the ways in which Gypsy Travellers' movement, both spatial and metaphoric, impacts upon their relation to space and place. Further, it will investigate how such relations deviate from dominant, metropolitan understandings of space and interrogate the ways these novel spatial conceptualizations challenge and resist sedentarist approaches. The application of Foucault's notion of heterotopia will be adopted to illuminate understandings of Gypsy Travellers' places and spaces as sites of resistance to the dominant spatial logic.

To examine representations and relations to space and places, sites and dwellings, this chapter will draw on interviews[9] I conducted with Gypsy Travellers in Scotland and England. I will also draw on the 2009 memoir *Gypsy Boy* by the Gypsy Traveller author Mikey Walsh (a pseudonym). The different spatial representations which emerge from the autobiography – 'an irresistible guide through this secret world',[10] as the *The New York Times* book review suggests – and from the ethnographic material provide a wide range of examples to explore the issues of space and mobility addressed in this chapter.[11]

Gypsy Travellers Mobility

> But the nomad is not necessarily one who moves: some voyages take place *in situ*, are trips in intensity[12]

The negative construction of Gypsy Travellers' mobility is 'deeply rooted in Western thinking where movement has traditionally been considered as something other than the norm'; mobility invades the settled and sedentary way of life and represents a threat with the potential to 'transgress existing power structures'.[13] The fear of 'shifting' people has to be understood in the context of a sedentary, progress-oriented metaphysics whereby the itinerant way of life is seen as almost 'primitive', as an earlier stage of humanity.[14]

The wealth of literature which (mis)represents Gypsy Travellers is testimony to an internalized and consolidated misunderstanding of nomadism and nomads. Judith Okely's seminal work on Gypsies and Travellers in Britain points out countless folk tales and children stories where the Gypsy is scorned, mocked, feared and demonized.[15] The seed of an anti-nomadic logic is instilled in a childhood imaginary and inhabits collective consciousness from early days. Nomadism has been and still is profoundly misunderstood; this failure affects both the settled, dominant society – ravaged by fear and obsessed by the mission to regulate and house its 'vagrants' – and the Gypsy Travellers whose identities and ways of life are constantly under threat.

From a sedentary perspective, nomadism is about moving, about routes rather than roots. However, 'nomadism is a state of mind rather than a state of action',[16] as Kenrick and Clark remind us; it is not only about corporeal travel, but also about emotional and relational mobility. 'The mobility of Gypsy Travellers involves the transmission of objects, expressions of support, the creation of landscapes of memories, as well as physical and emotional returns to particular places'.[17] The mere act of telling 'moving on' tales and singing songs about travel and movement relates to spatial travel as well as to emotional travel, as an act of coming together in the process of sharing social practices and customs. As Shubin observes:

> for Gypsy Travellers, mobility in itself is a fluid and transformative process which involves anticipating movement and adapting to changing living conditions with the possibility of travel in mind. These emotional, symbolic and imagined aspects that accompany the physical movement of Travellers are reflected in maintaining the travelling 'atmosphere' and customs through religious meetings and festivals, which have taken the place of traditional Gypsy Traveller gatherings.[18]

The emotional and physical nature of travel entails both dislocation and displacement; metaphorical and physical movements are realized across a set of complex interrelationships. These itinerant spatial and social practices defy hegemonic understandings of mobility and remain profoundly misunderstood, essentialized and often oppressed. Hence, in dominant discourse Gypsy Travellers are seen as 'the quintessential "others", living amidst sedentary populations but maintaining a stubborn commitment to a separate, travelling lifestyle. Implicit in this discourse is the assumption that this separateness entails a denial of the responsibilities of citizenship'.[19]

Britain has a longstanding history of policy and practice aimed at curbing mobility.[20] For instance, as Shubin observes with reference to the Scottish context, the 1984 Roads Scotland Act and the 1986 Public Order Act 'demonstrate understandings of physical travel as chaotic and disordered and as something that must be brought under control'.[21] This legislation forbids encampments and campfires by a road, prevents gatherings of more than twenty people, imposes boundaries, limits travel and ultimately criminalizes most aspects of the Gypsy Travellers' ways of live. Based on familial interdependence, Travellers usually move and pitch up in groups to both benefit from the support of the extended family as well as for work reasons. For instance, care for the elderly and for children is often shared across the family; as Maria, a young Gypsy Traveller I interviewed, explained:

> Here you always know that there is people to help out with the elderly people, like my granny. Or with the children – somebody who has got three children and wants to go to the shops for an hour could ask the neighbours to look after them. It's normal. It's what we are used to.[22]

Despite globalization's promise of hypermobility, the limitations imposed by coercive policies severely affect these practices and threaten Gypsy Travellers' social structures. Lol, a Gypsy Traveller woman, explained to me the motives behind movement:

> Why people like to move: number one, nine times out of ten is to do with work, but also we are going to pull with such and nobody, we go every year; you can go to your

family, you can go to your children, you can go back to your parents. And basically is to do with being able to live the lifestyle that you have been brought up with.[23]

This description of mobility clearly spells out economic factors as well as social factors – interdependence across the community – both elements which are at the heart of nomadic ways of life. However, aggregation, cooperation and inter-relations have been and still are targeted by sedentarist laws. While policies may differ across countries, there exists a consistent trend across the West.[24] Curbing nomadism, in these specific examples, is associated with improved security, and ultimately with offering nomads 'better' opportunities, to impose sedentarism as the norm and to restrain – to different degrees – any form of nomadism. Areas of policy which affect nomadic lifestyles are diverse and range from healthcare to education, from social services to accommodation and work.

Policy reduces nomadism to movement and obliterates vital aspects of nomadic life: 'nomadism involves much more than mobility, including valuing the tradition or even potential of nomadism, economic independence and flexibility, different family structure, language, and caravan dwelling. Instead, the universal nomad is seen as the quintessential mover'.[25] The image of the wandering, itinerant and ever-shifting Gypsy only populates peoples' imaginations; as Deleuze and Guattari provocatively ask, 'even if the journey is a motionless one, even if it occurs on the spot, imperceptible, unexpected, and subterranean, we must ask ourselves, "Who are our nomads today?"'.[26] In Britain, as a result of restrictive policies, Gypsy Travellers retain their lifestyle in different ways. Some are housed, and partially 'settled', 'many now live in grim government encampments on the outskirts of urban areas … others have integrated into dominant settled society'.[27]

Today's nomads perhaps only move for a couple of weeks a year for vacation, they might be 'settled' in the same council site and might fear moving on for lack of authorized sites. The shortage of 'legal'[28] sites in which to camp and the proliferation of unauthorized encampments have become a key concern in the last decades in Britain and – while the discussion on policies goes beyond the scope of my study – it is imperative to point out this anxiety over space. In Foucault's 'Of Other Spaces' we read that 'the present epoch will perhaps be above all the epoch of space';[29] Lefebvre would echo his words a few years later: 'space has now become one of the new "scarcities"'.[30] The relevance of these spatial concerns to the Gypsy Travellers' communities could not be more uncannily fitting, evoking the terrifying images of October 2011 – when Dale Farm, the largest Gypsy and Travellers site in Europe, was forcibly dismantled by the police. The brutal eviction of over five hundred people is described by Imogen Tyler as 'one of the most disturbing and corrosive events in the recent history of British race relations, the consequences of which are still unfolding'.[31] The sedentarist struggle to curb nomadism is an effective embodiment of the connections between space and power. The contestation about space and about the ways space is inhabited, lived and experienced exemplifies the fraught relations between settled and nomad communities, centre and margin, norm and deviation.

Deterritorialized Spaces and Heterotopias of Deviation

In *A Thousand Plateaus* (1987) Deleuze and Guattari discuss nomadism as a social practice as well as a spatial practice. In differentiating the nomad from the migrant, they introduce the metaphor of the trajectory; rather than just going from point to point out of necessity, nomads are concerned with the trajectory whereby the route, the trail is crucial, it is in itself an objective.[32]

> [E]ven though the nomadic trajectory may follow trails or customary routes, it does not fulfill the function of the sedentary road, which is to *parcel out a closed space to people,* assigning each person a share and regulating the communication between shares. The nomadic trajectory does the opposite: it *distributes people (or animals) in an open space,* one that is indefinite and noncommunicating ... sedentary space is striated, by walls, enclosures, and roads between enclosures, while nomad space is smooth, marked only by 'traits' that are effaced and displaced with the trajectory.[33]

The nomadic trajectory defies the sedentarist spatial logic: parcelling out is supplanted by distributing, closed space becomes open space. The striated, enclosed sedentary space cannot conceive of the smooth, open nomadic space; the transient traces along a trajectory are a threat to normative topographic practices, to the permanent demarcation and delimitations of sedentarist space.

'The nomad, nomad space, is localized and not delimited',[34] it entails a rupture from normative space, it is a deviation. In his first principle of heterotopia, Foucault theorizes 'heterotopias of deviation: those places in which individuals whose behaviour is deviant in relation to the required mean or norm are placed'.[35] Gypsy Travellers' ways of life differ from the norm, 'in constantly redrawing boundaries, Gypsies are perceived by the dominant society as "deviant"',[36] their sites and encampments are *other spaces* for they do not fit into established social (and spatial) order. 'Gypsy-Travellers, especially those who still pursue a nomadic way of life, violate the sedentarist basis of law in modernity because of their different approach to spatial ordering and management'.[37]

The otherness which such sites exude has nothing to do with orientalist conceptions of nomadism, rather it is the result of a complex interplay of exclusion and self-exclusion, marginalization and self-preservation. Such dynamics and power struggles all take place over space. As Levinson and Sparkes observe, most Gypsy Travellers exist on the peripheries, they occupy places at the margin, they often dwell in spaces which 'have been rejected by the dominant society'.[38] Whether on an authorized or unauthorized encampment, Gypsy Travellers' sites tend to be spaces discarded by society, 'those marvellous empty zones at the edge of cities'[39] as Foucault would put it. Doron points out that

> most of these *terrains vagues* have been populated by marginalized communities ... These spaces are difficult to utilize by the common means of planning and architecture for various reasons: they might be physically demanding, not easily accessible, too small or of irregular shape, with tricky ownership rights, not lucrative, with other regular usage at some part of the day that might be in discord with other suggested usages, and so on.[40]

Functioning like what Doron terms 'dead zones',[41] these *other spaces* are left empty by dominant, sedentarist topography and are in fact encampments (both authorized and unauthorized) inhabited by Gypsy Travellers. The memoir *Gypsy Boy* provides rich and insightful representations of spatiality in this context; the first person narrator tells of the nature and ubiquity of such sites:

> Gypsy encampments are everywhere. Most are secluded, hidden away down inconspicuous back roads ... Our next campsite was in a dirty little town, through a dirty little road and up behind a dirty old petrol station, where we were surrounded by several overgrown fields filled with rubbish.[42]

The notion of waste and refuse pervades this image; the heaps of rubbish surrounding the site function as a metaphoric transposition of nomadism in the sedentarist spatial ordering. The human and spatial waste, to borrow from Bauman, are both hidden away from sight, secluded, othered. The space which is left deterritorialized, the non-space, the *other space* is re-territorialized by Gypsy Travellers' encampments; as Deleuze and Guattari have it, 'the nomad reterritorializes on deterritorialization itself'.[43] Upon arrival on a new site, Mikey, the narrator in *Gypsy Boy* recounts:

> After the darkness of the forest, the light above the clearing shone through so brightly that our eyes had to adjust ... The clearing was like a huge swamp. Not a blade of grass, not a tree in sight, but endless mud and water and several towering pillars, with iron steps at the sides and thick electric cables balancing from the top ... Welcome Travellers to Warren Woods Caravan Park.[44]

Warren Woods Caravan Park in this description is a treeless, mud-filled swamp, devoid of the very idea of a 'park'. Here the space is reterritorialized to become *another* space. For the nomad 'the land ceases to be land, tending to become simply ground *(sol)* or support'[45] to another spatial ordering, both imposed and self-designated. This spatial dimension exemplifies the Foucauldian notion of heterotopia in multifarious ways. As Cenzatti reminds us,

> modern heterotopias, then, are 'other spaces' on the one hand because they are made other by the top-down making of places of exclusion; on the other hand, they are made other by the deviant groups that live in and appropriate those places.[46]

The marginal sites at the edge of cities, the non-places, the *terrains vagues* or dead zones are spaces which, as excess or refuse within the sedentarist normative topography, are designated as *other places*. This is a top-down production of places of exclusion. In Britain, the formula of City Council-managed encampments enhances this formalized fabrication and vigilance of places of exclusion, of other places. On the other hand, privately owned encampments (often by Gypsy Travellers themselves) as well as unauthorized ones represent forms of appropriated *other spaces*, reterritorialized places which the norm has deterritorialized:

> Coming back to the camp from anywhere else was like entering into another world: a full-scale exotic trailer-filled town, created and built by Gypsies for Gypsies. Fresh concrete had been poured on top of the mud that had once been everywhere, and a smart road of jet-black tarmac flowed right through it. At the main entrance the walls curved and spiralled ingeniously like frozen waves. At the very tip of each solid wave stood the life-sized stone head of a wild horse, peering like a milky-eyed guardian at the people passing below. And inside, the plots were no longer marked out with red strings, but with scarlet brick walls, eight feet tall, surrounding each home like gigantic theatrical curtains.[47]

This extract from *Gypsy Boy* illustrates heterotopia at its fullest. The camp is described like *another* world created by Gypsies and for Gypsies, a space produced by virtue of top-down exclusion and ultimately re-appropriated to become another place. Gypsy Travellers' agency is here exemplified by the spectacular re-making of this swamp into something else, something other. According to Foucault's third principle

> heterotopia is capable of juxtaposing in a single real place several spaces, several sites that are in themselves incompatible. Thus it is that the theater brings onto the rectangle of the stage, one after the other, a whole series of places that are foreign to one another.[48]

Mikey's dramatic description of this camp powerfully invokes this principle. Like a theater stage, this dead zone is turned into a full-scale town, mud replaced by concrete, a shiny tarmac road flowing through it. The juxtaposition of incompatible sites on one real place is uncanny: a discarded plot of land, secluded from sight in the middle of nowhere hosts curved walls 'like frozen waves', and on each wave towered 'the life-sized stone head of a wild horse'. The tall, 'scarlet' brick walls close off this 'exotic' town like 'gigantic theatrical curtains', this is after all a spectacle not to be seen. While the 'smart' road at the heart of the site functions as both metaphor and reminder of spatial and imaginary travel, the horse – also attached to the Gypsy imaginary – seems to incarnate at once self-protection and self-surveillance like a 'guardian'. The contested and fraught power relations over such sites are both implied in this representation as well as challenged. The overt heterotopic nature of this site, its *other* dimension, opens up further lines of inquiry into the application of heterotopology to both nomadism as a social practice as well as nomadic sites and dwellings.

Visibility and Resistance

Heidi Sohn argues that, 'an exception to uniformity and homogeneity, heterotopia opens up pathways for the deconstruction of sameness and its subversion, becoming the antidote against the erasure of difference implicit in the progression of the cultural logic of late capitalism'.[49] Heterotopia, both etymologically and conceptually, entails difference as a resistance to the norm (sameness) as well as a venture to protect heterogeneity against a homogenizing progressive logic. Gypsy Travellers' heterotopic sites pose a challenge to the settled communities, but most importantly seek to defy the cultural erosion which is inevitably caused by and is inherent in the sedentarist project. Sedentarism's monopolized forms of power constrain nomadism – hence their heterotopic emplacements are both a result of such oppression as well as a response to it. Sohn's description of heterotopias as an 'antidote against the erasure of difference' aptly fits Gypsy Travellers sites whereby nomadic ways of lives are practiced, preserved and handed down to generations, both challenging sedentarism and protecting their endangered cultural difference.

According to the fifth principle, 'heterotopias always presuppose a system of opening and closing that both isolates them and makes them penetrable. In general heterotopic space is not freely accessible like a public space'.[50] Mikey's description of the encampment 'by Gypsies and for Gypsies' is unmistakably conjured up in this principle. Gypsy Travellers' sites are founded on a strict code of access, of inclusion and exclusion. The importance of keeping a nomadic way of life, either with spatial or imaginary travel, is key; however, this can only be achieved as a group. Gypsy Travellers' social structures are based on interdependence and interrelations, thus an encampment does not necessarily mean a conglomerate of Gypsy Travellers living in proximity. As Lol said, 'we are going to pull with such and nobody'; this statement implies a process of selection based on a wide range of criteria. Indeed, family relations and interrelations strictly regulate spatial ordering[51] and everybody is careful about the ways in which space is collectively territorialized. Hence, access on a site, whether it is authorized or unauthorized, is carefully regulated and monitored; Gypsy Travellers' sites, in truly heterotopic vein, are 'not freely accessible like a public space'. Insider-outsider relations are extremely complex and extend beyond space. As Karner reminds us, Judith Okely

> shows that the Travellers' cultural logic keeps the categories of 'self' and 'other' strictly separate. The insides of camps, trailers, and bodies all symbolize the 'ethnic self', which must be kept separate from, and uncontaminated by, the 'outside' (symbolic of sedentary/Gorgio society). Every 'crossing' or blurring of the self-other/in-side-outside boundary is a source of pollution that must be guarded against – hence the pronounced preference for endogamy and Travellers' rituals of cleanliness.[52]

This cultural logic based on separation, and the enhanced preoccupation with cleanliness – both physical and metaphorical – not only govern social relations, but also dictate entry and access to Gypsy Travellers' spatial realities. Hence, screening away from the outside, so spectacularly realized on the site described by Mikey in *Gypsy Boy,* is a concern and a priority for many Gypsy Travellers. Lol, who lives in a mobile home, explained to me the travelling way of life: 'It's not easy, it's a hard life, but it's a clean life … All this low life is creeping in from outside'. Here the metaphor of cleanliness is eloquent; at once it illuminates this cultural logic and reinforces the heterotopic nature of such sites. Greenfields and Smith observe that 'the association of outsider groups with dirt and unhygienic practices is a universal characteristic of insider-outsider relations'[53] in the Gypsy Travellers context. Lol's reference to 'low life' which creeps in, signifies anxiety and preoccupation with non travelling ways of life (sedentary) penetrating in her world. In this striking example the 'in' and the 'outside' are unmistakably put in an antithetical relation governed by a metaphorical cleanliness.

Referring to other Travellers who adopted a sedentary life, Lol said: 'When they do seem to settle down, they get bad habits, they mingle too much with the Gorgias[54] and they got Gorgias' ways … and say words that are no no to us. And they lose it … whatever they have had'. It is interesting to note that settling down, leaving the 'inside' to inhabit the 'outside', entails an actual loss – once tainted by unclean ways of life of the outer world, it is impossible to return, to make it up. As Levinson and Sparkes observed in their ethnographic study about Gypsy Travellers in Britain, 'the simple fact remains that contact with non-Gypsies for many in this study entails the risk of pollution'.[55]

The inside-outside dichotomy undeniably constructs barriers which shield travelling life from the outside (sedentary) world, and contributes to the solidification of stereotypes about travelling communities as being mysterious, secretive, unknown. This is also another aspect of heterotopia; for Soja, heterotopias are spaces 'linked to the clandestine or underground side of social life' which imply 'a partial unknowability … mystery and secretiveness.[56] The dominant discourse, reinvigorated of late by a Channel 4 pseudo-documentary about travelling communities in the UK,[57] insists on secrecy as both a badge of honour and shame for Gypsy Travellers. Reproducing stale stereotypes about the mysterious and exotic Gypsy figure, such discourse both titillates undue curiosities about these communities as well as justifies its mission of uncovering the veiled truth of these 'secretive' people.

The voyeuristic gaze into Gypsy Travellers' lives granted by the media is coupled with closer surveillance – the state is increasingly more concerned about counting, ordering and categorizing its nomads. As a result, the invisibility of Gypsy Travellers communities – relegated to the social and topographic margins of metropolitan society – is both produced by dominant power as well as questioned by it. Imogen Tyler reminds us of Papadopoulos *et al.*'s notion of 'becoming imperceptible'[58] as 'the most effective tool that marginal populations can employ to oppose prevailing forms of geopolitical power. Certainly invisibility is an important strategy of evasion'.[59] Gypsy Travellers' invisibility – which metaphorically envelops their sites behind theatre curtains as Mikey recounts

– though being the product of exclusion, is indeed a strategy to defy geopolitical power. The erosion of this 'imperceptibility' triggered both by media exposure as well by ever more stringent sedentarist policies, is strongly perceived and criticized by Gypsy Travellers. Lol said:

> Everything is reporting, everything is the police, nanny state, very very bad, and is closing in, closing in, closing in. It's big brother, they want to know where you are, who you are, how many people there is here, how many people is there. I mean there is nothing more revolting than being somewhere and you have got to tell them how many people lives in your home. If I want 20 people in here, I don't want to have to tell the council.[60]

This is a sharp criticism of sovereignty and surveillance: police and state – suitably conceptualized as the eye of a reality show camera – are seen as an approaching force threatening their ways of life. The image of closing here is very effective as it recalls Deleuze and Guattari's analysis of sedentary space as striated and 'closed', in antithesis to smooth and 'open' nomadic space. Moreover, this passage also raises important questions about sedentarist policies and surveillance; once again word choice here is key. 'Revolting' has been theorized by Imogen Tyler in her recent book *Revolting Subjects* (2013) with reference to those who are both rendered *abject* by governmentality and who represent a precarious counter-public revolting against coercive ideologies.[61] Despite being among the 'revolting' groups in Tyler's analysis,[62] one of the disposable 'wasted humans', Lol, the Gypsy Traveller woman I interviewed, perceives the abject from the other end of the telescope: in her words the state's surveillance practices are 'revolting'. The state has rendered nomadic life untenable, and the renewed visibility and increased categorization of Gypsy Travellers have 'led to the immobilization within systems of bureaucracy and penal control'.[63] Amalia, a middle-aged Gypsy Traveller woman, said, 'we are what we are. We are a moving on people. They have to let people be independent. They pushed us *out* for centuries and all of a sudden they want to throw us *in*'. This plea for independence clearly draws out the terms of a century old power struggle where margins (out) and centre (in) are in constant tension. Infantilized by dominant power structures, nomads today are ever more visible, yet still dwell at the margins, and are increasingly more immobile.

This question of visibility and marginality, imperceptibility and recognition – as irreconcilable aspects of the same site – pertains to heterotopias of difference. Cenzatti points out that:

> these other spaces can never be fully understood, since we cannot know the 'other' and the group-specific cultures, codes, interactions and the 'unknowable and secretive' spaces the 'other' produces. Yet, to what extent heterotopias are visible and can be known depends on their position shifting between invisibility, marginalization, assertion of difference.[64]

Cenzatti's reading of Foucault's heterotopia, while not referring to any specific place, seems to gesture towards sites such as Gypsy Travellers' encampments. Their otherness, unknowability and secretiveness aptly fit this context; moreover, the shifting between invisibility and recognition summarizes the complex dynamics of exclusion and resistance which are at play in the power struggle over space. I argue that Gypsy Travellers' sites do not embody deviancy per se, but constitute forms of resistance to power structures; theirs is an empowered otherness. Interestingly, Kendall identifies Travellers' home places as 'sites of resistance' for the 'cultural survival' of marginal groups;[65] however, beyond self-preservation of certain cultural practices, such sites also subvert hegemony. Against the metropolitan, sedentarist logic, Amalia said: 'Nobody wants anything done. You have a problem because you made us a problem. We want to be left to do what we want, ourselves. People want to help themselves, be allowed to help themselves'. Here it emerges a cogent counter-argument: the 'burden' of settling nomads heralded by the state as a benevolent mission towards marginal communities is unmasked, rejected and challenged by Amalia. An assertion of individual and collective agency comes to the fore with clarity and force. Karner also discusses Gypsy Travellers' practices as counter-ideologies, claiming that 'their lived rejection of dominant values amounts to, in Roland Barthes's (1993) terminology, a powerful "de-naturalisation" of (post-)modern "common sense" and hence presents an ideological challenge for (post-)industrial societies'.[66]

Gypsy Travellers' social and spatial practices pose a real challenge to dominant, sedentarist – or as Karner has it – (post-)industrial societies. The emancipatory metaphor of nomadism, elaborated by de Certeau, Baudrillard, Deleuze and Guattari among others, interprets mobility as an escape from spatial order, as freedom from rules and regulations;[67] such a metaphor is valuable to further illuminate our discussion on resistance, though I do not wish to romanticize nomadism as an escape from both social participation and social obligations. Rather, my analysis seeks to deploy heterotopology in order to examine Gypsy Travellers' response to centralized territorialization and capitalist spatiality. For Lefebvre, heterotopia 'delineates liminal social spaces of possibility where "some thing different" is not only possible, but foundational for the defining of revolutionary trajectories'.[68] The difference as possibility and potential toward revolutionary trajectories is a fruitful way to approach Gypsy Travellers' spatiality. Lol said: 'I am not tied down. Being a Gypsy Traveller is a way of thinking'; this statement asserts difference and at the same time defies sedentarism at its very root. Sedentarist policies cannot ultimately change her travelling ways of life – since it is more than anything else a state of mind.

The ways Gypsy Travellers relate to and conceptualize time are also crucial to explore the notion of resistance. Foucault discusses heterotopias linked 'to time in its most fleeting, transitory, precarious aspect ... These heterotopies are not oriented toward the eternal, they are rather absolutely temporal'.⁶⁹ Part of the fourth principle of heterotopias, these heterochronies produce another mode of ordering time in space. Gypsy Travellers' spatial logic not only defies normative spatialities, it also challenges established understanding of time. Gypsy Travellers' temporalities are anchored in the transient, in the precarious; 'Travellers do indeed "manipulate time" – through their alternative work routine that resists sedentary ("proletarianising") control and, if nomadic, by deciding when to travel and how long to stay'.⁷⁰ The transient nature of Gypsy Travellers' sites is due to the omnipresent possibility of moving on – as Maria told me, to 'pick it [the mobile home] up and go' – thus quickly altering the topography of an encampment. Mikey Walsh recounts how, within hours his family shifted from one site to another, from one mobile home to a new one:

> We had been in the bungalow for just over a year when our father arrived home one day with the news that a new Gypsy camp was being built a few miles from where his family was living. He had bought a plot, a brand-new trailer and a new lorry to ship us all there ... And so we packed up and moved to start a new life just a few miles from Tory Manor, in West Sussex.⁷¹

Mikey's family departure altered the topography and the spatial relations on both sites – the one they left and the one they moved on to. The transitory nature of these spatialities is emphasized by the fleeting temporalities that govern them: 'Slowly the camp was taking shape ... Each of the families found their spot and within a few minutes the legs were wound down on the trailers and the dogs set free from the backs of the lorries'.⁷² This image of trailers' legs stretched into the ground – within a few minutes – and of the site taking shape almost instantaneously represents a rupture from sedentary temporalities, whereby such transience is only conceived in the context of festivals, holiday camping, and fairgrounds. The fluidity of space and time in the nomadic logic entails resistance to normative spatio-temporal dimensions, the delimited and universalized segments of time and space that sedentarism produces. To recall Soja, such localized spatialities and infinite temporalities trace new possibilities for alternative, revolutionary directions, *other* ways to inhabit space and time.

Sacred Spaces and Mobile Homes

Nomadic spatiality can be better understood in relation to what Foucault calls sacred spaces:

> And perhaps our life is still governed by a certain number of oppositions that remain inviolable, that our institutions and practices have not yet dared to break down. These are oppositions that we regard as simple givens: for example between private space and public space, between family space and social space, between cultural space and

useful space, between the space of leisure and that of work. All these are still nurtured by the hidden presence of the sacred.[73]

Gypsy Travellers' sites are desanctified spaces as their logic eschews these dichotomous oppositions. For instance, there is no work-leisure distinction,[74] and often 'sites constitute workplaces as well as home places';[75] further, the opposition of public and private space is debunked and complicated by the interrelational logic which pervades Gypsy Travellers' social structures. 'The doors seem seldom closed'[76] thus blurring the division between private and public space. Maria told me: 'When we go away, we never lock the door. We don't lock the door at night. We don't lock the door if we go to the shop do we? We leave all the windows open all the time'. On a similar note, an elderly woman, Margaret, said: 'I do the washing outside, the cooking outside. It's a different way of life'; Lol added: 'most people on a site have their things outside, they have got their washing machine outside, their dryer outside. I spend more time outside this place [on the site] than I do in [the trailer]'. Many daily tasks are performed in the open air on the site, like washing up, cooking, child-minding; such practices represent a rupture from sedentarist spatial divisions.

The family space/social space dichotomy is also mostly irrelevant to Gypsy Travellers; their familial and social structures defy such division. The extended family is key in Travellers' relations – for both work and every day life; thus, living in proximity, often on the same site, renders the familial-social division redundant. Talking about living on a site and engaging with neighbours, Maria explained to me:

> You could walk round the corner and see somebody familiar to yourself. As long as there is Travelling people around me ... There is always somebody at the window or somebody you can say hello to. That's why I like it better here. Somebody just comes in and have a cup of tea. I wouldn't have it any other way. There is always somebody stopping over. Because everybody knows one another. It is all very close-knit. Even people you don't know, you know of them.[77]

Lol said: 'If there is a slight chance you don't know some of them or their breed, before you turn around you know all of their family, their children, their husbands, their wives'. The ubiquitous presence of familiarity is enhanced by the fact that knowing others or knowing of others is based on interrelations – those who are not related to you but belong to a respected Traveller family automatically become part of the circle. Also, it is common practice among Gypsy Travellers to address an elder or somebody older than yourself as 'uncle' or 'aunty' – this custom, based on respect and reverence, effectively extends and blurs the familial over and into the social.

Foucault's heterotopology also includes a discussion of the relations between heterotopias and 'all the space that remains ... Either their role is to create a space of illusion that exposes every real space, all the sites inside of which human life is partitioned, as still more illusory'.[78] Gypsy Travellers' sites, as heterotopias, offer alternative spatialities which both challenge and expose all other sites where life is 'partitioned', delimited and enclosed, revealing their illusory nature. Such heterotopias create 'another real place, as perfect, as meticulous, as well arranged as ours is messy, ill-constructed, and jumbled. This latter type would be the heterotopia of compensation'.[79] Gypsy Travellers' dwellings – trailers, chalets and other types of mobile home – compensate the brutal enclosure of capitalist space and create *other places* which are perfect, meticulous, well-arranged places. Mikey Walsh's account of trailers seems to encapsulate heightened forms of compensation:

> Their trailers were monstrosities, created to mimic miniature palaces. Garish, flamboyant and overtly camp, we couldn't move for polished steel, mirrored cabinets and chrome. Every surface was carved from white, polished timber with a mirrored effect, and not one cupboard was without a glass window, so that the woman of the house could display her Crown Derby.[80]

Like small palaces, trailers are fashioned as perfect, well arranged places – often anthropomorphized; these dwellings are both home and transport and define living spaces. Within the emotional and physical nomadic spaces, mobile homes represent perfect 'real places'.

There is a plethora of literature on Gypsy Travellers' relations to mobile homes – especially about those who are in effect 'settled' (moving only for vacation) but still prefer mobile dwellings to a brick and mortar home. In *Gypsy Boy* Mikey recalls the story of his grandfather who had been housed among brick walls: 'but after three days Granddad, miserable in a home that didn't have wheels, refused to live in it any longer'.[81] The wheel-less home instilled misery in the old man. Indeed, in their ethnographic study Levinson and Sparkes observed a

> dislike of houses, despite the extra facilities, and a preference for caravans or trailers, on both physical grounds and because of connotations. Physically, it seems, increased space can feel like a loss of space ... Houses are associated with constraints and loss of freedom.[82]

Many of the women I interviewed were very clear about the set of spatial and emotional relations surrounding mobile homes:

> Lol: 'In a chalet you have the comfort of a house, you have the work around you, and you have to the people around you.'[83]

> Maria: 'I wouldn't like to be stuck in a house ... It's to do with being together'.[84]

> Margaret: 'In a house it will be too lonely. Here there is somebody you can turn to. In a house you feel shut off. The chalet is good; it is in between house and trailer. It's a good in-between medium. You are surrounded by bricks and walls in a house ... I couldn't be in a house, I would miss my relatives. I would feel lonely. In a house you are too cut off ... You can move these [chalet], you cannot move a house.'[85]

From these extracts mobile homes are perceived and experienced as both safe havens and places of relations, they are sites which enable travel, both spatial as well as imaginary. As Lol said, one day you might want to think 'hang on, I have had enough of this, let's pack up and shift'. Sensory deprivation characterizes Gypsy Travellers' perceptions of brick and mortar dwellings: 'feeling trapped; feeling cut off from social contact; a sense of dislocation from the past; feelings of claustrophobia'.[86]

Mobile homes exemplify the ultimate example of heterotopia; like trains – discussed by Foucault in 'of Other Spaces' – trailers, caravans and chalets are 'sites of transportation'; '(a train is an extraordinary bundle of relations because it is something through which one goes, it is also something by means of which one can go from one point to another, and then it is something that goes by)'.[87] Gypsy Travellers' mobile dwellings are both a home as well as a means of transportation, they defy dominant discursive categories and – in truly heterotopical terms – embody the juxtaposition of incompatible terms. Referring to her mobile home Margaret said: 'here you are in a long place ... Windows remind you of the outside, they remind you of the olden days singing and sitting outside. It is convenient'. The focus on shape ('a long place') in this spatial perception alludes to what Deleuze and Guattari have termed a 'nonlimited locality': the mobile home, with its long shape, aligns itself to the horizon and blends within it, thus failing to delimit space. Windows operate as both a reminder of the outside as well as of the past travelling life, when mobility was not curbed. Such dwelling hosts the potentials to being in other places at once – it conjugates in its shape home and travel, spatial and metaphoric movement. Foucault's essay 'Of Other Spaces' concludes with a heterotopological analysis of the boat.

> The boat is a floating piece of space, a place without a place, that exists by itself, that is closed in on itself and at the same time is given over to the infinity of the sea ... the boat has not only been for our civilization, from the sixteenth century until the present, the great instrument of economic development, but has been simultaneously the greatest reserve of the imagination. The ship is the heterotopia par excellence. The ship is the heterotopia *par excellence*. In civilizations without boats, dreams dry up, espionage takes the place of adventure, and police takes the place of pirates.[88]

This evocative image of the boat as heterotopia, as a place without a place, closed on itself and yet consigned to the boundlessness of open spaces, suitably fits the notion of mobile home. The ship is 'a richly ambivalent vessel ... it is an emplacement that is enclosed and yet open to the outside ... The ship not only visits different spaces, it reflects and incorporates them'.[89] Similarly, a mobile home visits different spaces and all the same encompasses all these sites within itself. Like the boat, a mobile home for Travellers is a 'reserve of the imagination', it involves the anticipation of movement, imaginary travel, it triggers memories of spatial travel and, by definition, it embodies the possibility of dislocation, as Deleuze and Guattari have it, some 'voyages take place *in situ*, are trips in intensity'.[90]

Foucault warns that 'in civilizations without boats, dreams dry up' – similarly, without mobile homes, Gypsy Travellers' imaginary (and corporeal) travel would cease, their associations with a traditional lifestyle would be severed. At the mere thought of being surrounded by brick walls Lol said: 'Once you are behind that door you are behind that door … It's a freedom thing'. Gypsy Travellers' mobility enables freedom – both spatial and imaginary, it offers the opportunity to set off different kinds of journeys. It interrogates and illuminates sedentary practices and proposes alternatives ways to inhabit and dwell in space.

Acknowledgements

I am grateful to Mary Hendry for introducing me to these wonderful heterotopic realities and for allowing me entry into these 'spaces'; to Lol and Amalia for their time and wise words, and to all the women I interviewed.

5 HETEROTOPIAS OF ILLNESS

Stella Bolaki

In *Lost Bodies*, Laura Tanner writes, 'Although recent cultural criticism explores the stresses and fissures created when an embodied subject negotiates economic, political, and geographical landscapes, the normative body such criticism posits is most often a body that is healthy, functional and stable'.[1] This essay explores the relationship between bodies and space as well as between space, power and resistance in the narratives of people who inhabit what could be called 'heterotopias of illness'. The relationship between space and power, which is central to Foucault's work, is, as this volume demonstrates, one of the most urgent questions of our time. The ongoing critical conversation as to whether heterotopias are spaces of freedom or normalization, prompted by Foucault's ambiguous conception of 'other spaces', is more alive than ever in our globalized age. This is an era where hyper-mobility and cosmopolitanism find a striking contrast in the restricted movements of refugees, illegal migrants and other people for whom movement is a luxury. Debates about the nature of 'other spaces' are also relevant to discussions of medical spaces which, echoing scholars of heterotopias, could be seen as either 'vulnerable and marginalized spaces'[2] or spaces of 'Other voices'.[3] This chapter productively adapts some of Foucault's distinctions in 'Of Other Spaces' to explore the heterotopian qualities of illness and of medical spaces such as the tuberculosis sanatorium, the cancer clinic and the dementia ward in first- and third-person illness narratives. Such narratives have garnered more attention in the last few decades and have provided a means for patients to reclaim their bodies and their stories from medical discourse.

Medical spaces like the tuberculosis sanatorium and the cancer clinic interrupt the continuity and normality of everyday ordinary space in the same way that illness signals a break or discontinuity with ordinary time. However, while my essay uses heterotopia as a conceptual tool to think about illness (or the 'other spaces' of illness) more generally, it also highlights the ways in which one's culture, both broadly and narrowly defined, impacts the lived experience of disease and of space, and how illness narratives may draw attention to health inequalities in diverse cultural, social, political and economic settings.[4] This is in line with recent concerns in disability studies and medical humanities that emphasize the need to pay attention to such differences and consider the aims of these fields in a global context.[5] The three illness narratives examined in this chapter – *Madonna Swan: A Lakota Woman's Story* (1991), Audre Lorde's 'A Burst of Light: Living with Cancer' (1988), and Linda Grant's *Remind Me Who I Am Again* (1998) – demonstrate that an exploration of space foregrounds issues of power and privilege which in turn connect to race, ethnicity and culture. Despite the differences in terms of the historical period, circumstances in which they were written, and conditions that they narrativise, the spaces of illness described in these accounts acquire an additional 'otherness' that turns them into *other* 'other spaces'.

Illness as Heterotopia

Illness is often conceived in spatial terms. In her 1926 essay 'On Being Ill' Virginia Woolf describes illness as 'an undiscovered country'.[6] Most well-known perhaps is Susan Sontag's metaphor of 'the kingdom of the well and the kingdom of the sick' where 'everyone holds dual citizenship' in *Illness as Metaphor* (1978): 'Although we all prefer to use only the good passport, sooner or later each of us is obliged, at least for a spell, to identify ourselves as citizens of *that other place*'.[7] The transitions of illness are marked by space-specific rituals involving receiving treatment, as in the case of chemotherapy for cancer and, in Arthur Frank's phrase, 'ceremonies of recovery' to mark its remission.[8] With the age of acute disease having come to an end in the Western world, many patients live in 'the remission society'[9] or what could be called 'chronic' heterotopias,[10] and are required to spend short or long periods in medical spaces, such as hospitals, clinics, or hospices, not to mention the medical waiting room. The latter in 'assigning its inhabitants a body that is symbolically – and often biologically – impaired'[11] is perhaps what resonates the most with the medical interpretation of the term heterotopia, namely 'a spatial displacement of normal tissue'.[12] This interpretation blurs the boundaries between normal and abnormal or healthy and unhealthy, just as occupying the medical waiting room uncovers a body's vulnerability irrespective of whether that body is healthy or not.

In 'Of Other Spaces' Foucault refers to the space which 'draws us out of ourselves'.[13] As Peter Johnson suggests, 'this is crucial. Heterotopias draw us out of ourselves in peculiar ways; they display and inaugurate a difference and challenge the space in which we may feel at home'.[14] Illness does something similar, drawing us out of our familiar bodies and spaces, temporarily or sometimes permanently. But as the discussion of the three narratives below demonstrates, the 'difference' that heterotopias display and inaugurate is not the same for all patients.

Heterotopia of Permanence

Madonna Swan's oral history, which includes her long struggle with tuberculosis during the nineteen-forties, was published only in 1991. The story of her life is told through Mark St. Pierre who, as we find out on the cover of *Madonna Swan: A Lakota Woman's Story*, 'has lived among the Lakota people since 1971, both as an educator and as an encourager of American Indian art'. While Swan recounts various events of her life on the Cheyenne River Sioux Reservation in South Dakota, the most unsettling parts of the book are those where she talks about her long confinement or the 'years of sad isolation',[15] as she calls them, in the Sioux Sanitorium run by the United States Public Health Service. When she was sixteen, Swan was removed from her everyday life and sent to a place from which she believed she might not return. Heterotopia can be described as not merely a space but rather a time-space. Lieven De Cauter and Michiel Dehaene explain that 'heterotopia is the counterpart of what an event is in time, an eruption, an apparition, an absolute discontinuity, taking on its heterotopian character at those times when the event in question is made *permanent* and translated into a specific architecture'.[16] The event in Swan's case is tuberculosis, which, as Lisa Diedrich argues citing Foucault's 1976 lectures, became 'endemic' in the Native American population after increased contact with whites: 'Death was no longer something that suddenly swooped down on life – as in an epidemic. Death was now something *permanent*, something that slips into life, perpetually gnaws at it, diminishes it and weakens it'.[17] This section explores how the sanatorium, a typical site for TB patients in the twentieth century, becomes 'a heterotopia of permanence' for Native Americans.

Foucault's discussion on the relation between heterotopias of crisis and heterotopias of deviation (the former, as he claims in 'Of Other Spaces', has been replaced by the latter in modern society) is relevant to Swan's account of her life as an adolescent girl and later a patient at the sanatorium. When she menstruates, she briefly moves into a heterotopian space fashioned or reserved for such a life change. The following description resonates with Victor Turner's idea of 'liminality', that is, a set of ritual cultural practices which mark different stages of human life or states of transition:

> Among the Lakota people we have a way of marking a ceremony for young women when they have their first moon, or monthly cycle. The day that I got my first moon, we were living outside Cherry Creek. Mom and Grandma closed off a little space for me to stay with ropes and blankets.[18]

Swan stayed in Grandma's cabin for four days and four nights. When she went back to her normal life, things were different, by which she means that her relations to men were governed by constraints. St Pierre explains in a note that, according to elders from Swan's community, confinement of the female in a distinct space did not serve to disgrace the woman but 'to isolate her female powers from male spiritual ones'. This indicated 'a tremendous respect for the sacred mystery surrounding female procreative powers'. Swan's narrative shows that heterotopias of crisis still exist in some form within Native American tribes: in the pre-reservation society, according to St. Pierre's note, menstruating women were isolated 'in a small ribbed structure somewhat like a sweat lodge, far away from the main lodge';[19] in Swan's case, an enclosure was made to fit that old pattern for the ceremony.

The brief stay in the cabin and the ceremonies of entering adulthood contrast starkly with Swan's later confinement in the Sioux Sanitorium, even though discipline is a characteristic shared by both spaces. When Swan finds out that she needs to be confined or else they will 'put red tags' on, that is quarantine, their house, she exclaims: 'I don't ever want to come home again, because I am a disgrace'.[20] Tuberculosis, which was cast as 'the romantic disease'[21] in the eighteenth and nineteenth centuries, became associated with poor and immigrant communities in the twentieth century, and TB patients and their families were treated as lepers. The stigma attached to the condition makes the sanatorium a heterotopia of deviation alongside other spaces discussed by Foucault, most notably prisons, asylums and psychiatric institutions.

Foucault writes that heterotopias 'presuppose a system of opening and closing that both isolates them and makes them penetrable'. They are 'not freely accessible like a public place. Either the entry is compulsory, as in the case of entering a barracks or a prison, or else the individual has to submit to rites and purifications'.[22] In her narrative, Swan describes vividly the boundaries between the realm of the healthy, going about their daily lives, and that of the sick who are condemned to passively watch such an envious spectacle:

> Living in the san would make you feel like an outcast with some filthy disease like leprosy. We couldn't go outside. We were allowed only to stand out on the little balconies and look out across to Rapid City, watching people go about their daily lives, enjoying life. From 1944, through 1945 and 1946, until 1947, we were not allowed to go outdoors, not stand on the ground, Maka Ina, Mother Earth, ever once.[23]

The Sioux Sanitarium resembles a prison in many ways. Patients are dressed alike 'like the inmates at the penitentiary in Sioux Falls, all dressed in stripes'. As Swan explains, 'I guess that was intended to keep us from escaping. If we could escape we could not get far. Of course, our regular clothes were locked'.[24] As in some prisons, the healthier patients have to work (for example, they bathe the sickest ones), and in case of insubordination patients get further restricted. There are also private rooms which Swan compares to death row – 'you knew you were a goner if you were put into one of them'[25] – while getting bad news on medical tests is described as 'seeing the parole board and always being told maybe next time'.[26]

Even though many critics make a distinction between the characteristics of heterotopias and the camp,[27] Swan's story – and the condition it narrativises – recalls what Giorgio Agamben calls 'bare life'. The sanatorium is a form of concentration camp or a liminal space between sanatorium and cemetery, the latter being a heterotopia that, for the individual, begins with an absolute 'heterochrony, the loss of life'.[28] When Swan sees her friends and relatives die one after another, she realizes that 'this was a place for people to die'.[29] Early on, she noted that patients were not separated according to the seriousness of their condition and were not treated with any proper medication apart from cod liver oil and the daily routine of the bean bags (which were supposed to collapse their lungs and kill the germ). Food is scarce and the standards of hygiene are poor. Most chilling of all is Swan's 'diary of death' which turns the book into a memorial of a whole community of people, mostly Native Americans, who were left to die:

> Death was the only way anybody left ... and there were many. I had a diary and in it I wrote down the little things that happened each day, things girls outside the san would never have bothered to write down ... I wrote down the names of those that had died. Bernice Long also kept a diary. These were a daily log for us. In 1950 we went through our diaries together. We counted five hundred deaths.[30]

Although heterotopias function on the basis of a mechanism of opening and closing, Foucault notes that some heterotopias 'hide curious exclusions. Everyone can enter into these heterotopic sites, but in fact that is only an illusion – we think we enter where we are, by the very fact that we enter, excluded'.[31] Foucault's example is American motel rooms, but, in Swan's case, her ambivalent insider-outsider status in relation to the sanatorium site is tied to persistent racial hierarchies. The sanatorium reserved exclusively for Native Americans can be seen as an extreme version of the tribal isolation enforced by the Native American reservation system. Swan's narrative, according to Diedrich, 'records the letting die mode of population control'.[32] The Sioux Sanatorium is a place where Native American patients *wait* to die, while the white sanatorium, Sanator – to which Swan gains admission after she manages to escape – is the opposite of the death camp. It can be compared to a type of refugee camp where normality (within the constraints of illness) is reinstated: 'Sanator was the "white" sanitorium for TB. Standing Bear's daughter was the first Indian in there, and she got well'.[33] As Swan notes with surprise when she enters this space, '[t]he grounds were pretty, with trees, flowers and all. The patients were walking the grounds with their own clothes on'. Despite other patients' racism towards Swan, the white doctor who treats her promises to do 'anything in [his] power' to help her recover in contrast to the Sioux Sanitorium doctor who had told her that she would most definitely die.[34] Ironically, Swan, one of the first people to undergo experimental surgery for tuberculosis in the United States, ends up saving a great amount of white people, the same people who were responsible for the decimation of the Native American population through such a 'white' disease in the first place.

Architecture of the Holiday

In her influential work drawing on her cancer experience, Audre Lorde offers a profound insight into the power dynamics of the doctor/patient relationship and the role of the healthcare institution, as well as of women's collusion in their 'infantilization' as patients. Unlike Swan whose narrative is framed by someone else, Lorde has full control of her story, and her adamant refusal of prosthesis after her mastectomy in 'The Cancer Journals' (1980) has created a powerful narrative of feminist and patient resistance. 'A Burst of Light', less known than 'The Cancer Journals', consists of journal entries written during the first three years of living with cancer of the liver (1984-7). A few of those entries detail her brief stay in a clinic, Lukas Klinik in Arlesheim, Switzerland where she sought alternative cancer treatment (following Rudolf Steiner's anthroposophical medicine).[35]

The diary entries immediately draw attention to this 'other space': its otherness emerges from a series of factors. First it is an unusual place for an American patient: '[h]ow does an American come to be at the Lukas Klinik in Switzerland? Americans are known to be quite provincial'.[36] This is the question other patients often ask Lorde. Elizabeth Alexander sees Lorde's decision not to follow the doctor's prescriptive path and instead travel to Europe as a metaphor for black struggle:

> I can't even imagine being Audre Lorde and the doctors saying you will die if we don't cut your liver out. And one after the other and saying you know, I'm going to – you know, there are other ways of thinking in the world and I'm going to go around the world. And I'm going to learn how other people have dealt with this. I think that's a metaphor, too. That's a metaphor. You know, what would it mean – what would it mean if all of the black women throughout history and to this day had swallowed and acted upon that which was said about us? We wouldn't survive. I don't think we would.[37]

However, the question 'how does an American come to be at the Lukas Klinik in Switzerland' bears additional weight for Lorde who is the only non-white patient in the clinic. Unlike Swan's experience, the clinic is a place Lorde chooses and is able to enter, but to which she responds with ambiguity: '[i]t's something different from narcotics and other terminal aids, which is all Dr. C. had to offer me in New York City'.[38] But in a later entry she is skeptical: '[t]here is a part of me that wants to dismiss everything here other than Iscador [complementary treatment for cancer] as irrelevant or at least not useful to me, even before I try it, but I think that is very narrow and counterproductive'.[39] This passage suggests that the clinic is not simply a place of health treatment but also a social microcosm with its own norms and regulations. As in 'The Cancer Journals', Lorde in particular reveals the ways in which the black and lesbian author-as-patient experienced not only exclusion and discrimination, but also compulsion to perform acceptable identities for her health providers.

Lorde finds the 'deep serenity' that characterizes the Lukas Klinik both 'relaxing' and 'uncomfortable'. Despite its luxury, the place has an austerity represented by the granite statue of Rudolf Steiner with its 'blunt and massive' look in front of the terraces and the single pastel decorating the walls of the patients' rooms. Details that Lorde mentions in her entries evoke the institutionalized character of the place: though she notes that 'we all wear our own clothes',[40] later she compares the atmosphere of the clinic to the sanatorium from Thomas Mann's *The Magic Mountain*. Nevertheless, Lorde finds 'the calm directness' of the doctors a welcome alternative to the reality of US hospitals.[41]

While she agrees with the anthroposophical medical approach that has a foundation in a spiritual-scientific understanding of the human being, Lorde objects to the rigid rules through which this approach is put into practice in the clinic: '[n]obody believes in talking about feelings, even the strong expression of which is considered to be harmful or at least stressful to be beneficial'.[42] For the author of 'The Transformation of Silence into Language and Action' and 'Uses of the Erotic', articulated and expressed feeling is a source of power that cannot be excluded from the therapeutic process. Lorde describes in particular the cardinal rules that apply during the communal meals: '[t]he feeling in the dining room is genteel, cultivated and totally formal'.[43] More importantly, she is critical of the fact that patients are not allowed to talk about their illness: 'I don't know what makes the anthroposophs think this sort of false socializing is not more stressful than expressing real feelings, but I find it terribly wearing. Mercifully, I can usually retreat behind the language barrier'.[44]

It is clear from the above descriptions of the clinic that heterotopias are 'spatial-cultural constellations'; they are not spaces or places per se but rather 'structures, systems or arrangements'.[45] In the name of health, and despite the freedom it offers Lorde from US invasive cancer procedures, the cancer clinic not only disciplines the sick body but also regulates behaviour and social interaction through particular routines or training, which bring to mind Foucault's discussion of modern power in *Discipline and Punish* (1975). Lorde's daily transgressions of the clinic regime are embraced despite the cost of isolation: 'In this place that makes such a point of togetherness and community, Frances [Lorde's partner at the time] and I sat through an ornate New Year's Eve dinner tonight surrounded by empty chairs on each side of us, an island unto ourselves in the festive hall'.[46]

For someone who has written so powerfully about difference and the importance of working together despite differences, the uniformity and sameness imposed upon the patients of the clinic is infuriating and flirts dangerously with racism. Early on Lorde has explained that she is the only woman of colour in the clinic with the exception of an East Indian woman who works there: 'I wonder how she feels as a woman of color among all these white ethnocentric Swiss ... I find her a touch of emotional color within a scene of extraordinary blandness'.[47] Colour often works symbolically in Lorde's work, as in the overwhelming and nauseous whiteness Lorde encounters in Jim-Crow Washington D.C (when a white waitress refuses to serve her family ice cream in the parlour) in her biomythography *Zami* (1982).[48] In 'A Burst of Light' Lorde describes a Christmas evening at the clinic with a dose of sarcasm: '[n]urses go around and open every door a bit so that everyone can hear the music. Soft lights shining in the twilight windows. Very lovely. Just don't be different. Don't even think about being different. It's bad for you.'[49]

The most evocative scene described in her journals from that time is when she criticizes the place's insularity:

> Good morning, Christmas. A Swiss bubble is keeping me from talking to my children and the women I love. The front desk won't put my calls through. Nobody here wants to pierce this fragile, delicate bubble that is the best of all possible worlds, they believe. So frighteningly insular. Don't they know good things get better by opening them up to others, giving and taking and changing? Most people here seem to feel that rigidity is a bona fide pathway to peace, and every fibre of me rebels against that.[50]

Lorde here objects to the practice of isolating ill people from the healthy and removing them from the so-called stresses of daily 'normal' life. This isolation may not be as extreme as in the case of Swan but in both cases it is a method of disciplining individual bodies in space. It also reinforces the stigma attached to illness and depoliticizes it by divorcing it from other everyday struggles that require, as Lorde argues, 'giving and taking and changing'.

It is interesting to compare Lorde's entry above with Barbara Ochsner's blog that features on the clinic's website. Barbara, mother and wife from Zürich, stayed at the Lukas Klinik in May and June 2009 for breast cancer treatment. Ochsner's entries allude to the heterotopian qualities of the clinic but seem to invest this other space primarily with a degree of freedom. Below are some excerpts from her blog:

Tuesday, 26 May 2009

In this small world of its own ... we have all the time in the world to ourselves.

Friday, 29 May 2009

With only a few mobile phone calls, I feel very comfortable to be without a schedule, radio or television. They are absolutely dispensable here. I have dived into the world within the walls of the clinic and into this new way of life. You can hardly imagine that my days flow by nevertheless quickly with so few must-haves and with so much time.

Saturday, 30 May 2009

What a luxury for a housewife like me just to sit at a richly decorated table three times a day without the shopping and the kitchen work ... In this sunny garden restaurant I feel as if I were on holidays.

Saturday, 6 June 2009

I laugh about myself and this unusual idleness. Here I can simply spoil myself.[51]

Through this juxtaposition, I do not wish to negate Ochsner's distinct experience of the clinic or suggest that every patient should turn their illness story into some form of public discourse and cultural critique like Lorde. Rather, I wish to illustrate how the same heterotopia (of course there are differences given the time gap between the two diaries) can be experienced differently by people. What Lorde describes as an insular 'bubble', Ochsner recasts as a refuge or sanctuary produced by physical boundaries and by a special discourse (the way the clinic's website describes or advertises this place, for instance) that allow her to feel simultaneously protected and free. The idea of the holiday conjured by Ochsner in her blog makes the clinic resemble a 'heterotopia of festivity' (albeit dark) – in a later entry Ochsner 'long[s]' for the day of her next check-up. The clinic could be compared to another heterotopia, described by Setha Low, namely 'a gated community'; a residential development 'surrounded by walls, fences or land' which 'separates its residents from the concerns of the world at large'.[52] While this kind of separation might be a necessary stage of recovery, Lorde's commitment to the politics of illness and to communicating a *black lesbian feminist* experience makes it impossible for her to enjoy such freedom. Just like the pink prosthesis she tries on at the hospital in 'The Cancer Journals', 'perching on her chest askew, awkwardly inert and lifeless ... having nothing to do with any me I could conceive of' and bearing 'the wrong color',[53] the clinic represents an *other* 'other space' for Lorde in its overall 'whiteness', rigidity and insularity.

Space for the Rest

Linda Grant's memoir *Remind Me Who I Am, Again* is an account of her mother's dementia and the hard decision to put her in a home as well as a story about second-generation Anglo-Jewry and cultural memory. The former allows Grant to explore moral dilemmas and social issues concerning the rights of old people but also the plight of families who are struck by dementia and remain silent due to its stigma. In 'Of Other Spaces' Foucault includes retirement homes under 'heterotopias of deviation' but notes that they can be situated on the borderline between the heterotopia of crisis and the heterotopia of deviation: 'old age is a crisis but it is also a deviation since in our society where leisure is the rule, idleness is a sort of deviation'.[54] Illness further exacerbates this dilemma. As Lucy Burke argues:

> Senility, now recast as Alzheimer's disease, has become a disease category, a pathology to be set apart from so-called healthy ageing. Old age becomes something benign but at a cost, as Cohen puts it, because 'those relegated to the victimhood of Alzheimer's now [have] to bear the dehumanizing brunt of a total and unquestionable pathology'.[55]

Grant's memoir explores the connection between power and space from various perspectives and specifically reflects on the Jewish Care centre where Rose becomes admitted after spending some time receiving care in her own home. Like other accounts of dementia, Grant exposes the cruel vagaries of health care in both Britain and America. She attacks the policy of 'Care in the Community' in the years of Reagan's and Thatcher's administration, which closed down mental hospitals and released inmates 'into the tender solicitude of the streets where we tourists could see many of them in New York, sleeping in doorways or on top of heating vents on icy days'.[56] 'Care in the Community' made it very difficult for any elderly person, including Grant's mother, to be admitted to a home paid by their local authority: 'If my mother had not been Jewish ... We would have trailed around many homes, examined the dried flowers arrangements and the antique clocks, walked among the gardens, inspected the kitchens, then demanded to see the books'.[57] Here 'the rites of purification', mentioned by Foucault when analyzing the intricate mechanism of opening and closing that applies to heterotopias, are undertaken by those who need to scrutinize the various homes available in order to find out the most suitable one.

Entrance into this kind of heterotopian space is, however, also policed in different ways: as Grant adds, '[w]e know also that you could inspect and demand and poke your nose in all you like and it wouldn't make any difference because when it came down to it, hardly anywhere actually accepts people with dementia'. As in the case of the white sanatorium that excludes Native Americans in *Madonna Swan*, many rest homes exclude people with dementia: '"No incontinents" is the "No coloureds, Jews or dogs" sign of the world of residential care', Grant explains. The Jewish Care centre which agrees to take Rose is then a kind of desegregated space, a heterotopia that functions as a sanctuary for both her daughters who are unable to cope with their mother's care anymore and the mother who is fortunate enough to 'fall into the hands of her own community'.[58] However, the centre is also a 'space for the rest, for what remains'[59] outside the normal within the confines of the rest home. This becomes clear when the two sisters visit another prominent Jewish institution in London with its own self-sufficient dementia unit that has 'its own dining room and television'. As they ask themselves, '[i]s it not the case that what it provides is a form of segregation, an apartheid separating the mad and the sane?'[60]

Grant is initially impressed by the existence of Jewish Care: 'There is something called Jewish Care, an entire social work organization dealing with all aspects of dependency – the elderly, the disabled, orphans, the unemployed, the blind, the mad and Holocaust survivors.'[61] A society can make a heterotopia function in a very different way in the course of history,[62] and Grant provides some factual information about the Jewish Care centre's past: '[it] had originally been a home for illegitimate babies, then taken over during the war to house poor Jews and subsequently became an old people's home'.[63] Once inside, the place does not stand up to Grant's expectations. Jewish Care, particularly its dementia unit, should be thought of as 'a heterotopia of compensation' in that it is meant to create a perfect or meticulous space for patients. However, Grant stresses its institutional character by comparing it to a hospital and a prison: 'There are chairs arranged in rows and a number of vinyl-topped tables to which the inmates shuffle a few feet for their meals ... This is the tiny prison in which my mother is going to spend the rest of her life'.[64]

The arrangement of chairs 'in a row' becomes a metonym for institutionalization and, as in the case of turberculosis, dementia is depicted in the memoir as a form of social death. When Grant sees her mother sitting in a row but not being unhappy about it, she realizes that she has lost her freedom, but corrects herself by adding that she lost that a long time ago. Rose's condition is always an obstacle when it comes to negotiating the authority and power of the space she is confined in. While she sees the care centre as 'a dungeon'[65] and refuses to accept it as her new home, she is unable to leave. Grant offers the following image of impotence: '[w]hen I pass on the bus, I see her, her chair facing the window and the traffic and the park. She's always staring out, talking to no one, like a prisoner who longs for the outside world but has forgotten how to escape'.[66] We are here a long way from Woolf's depiction of illness in 'On Being Ill' which allows patients to 'cease to be soldiers in the army of the upright', and which becomes an occasion for idleness, irresponsibility and playfulness.[67] Grant's portrayal of her mother's gradual institutionalization (let us not forget that Woolf is in bed but not in an institutional space) makes it difficult to see this heterotopia or illness itself as a welcome refuge from socially scripted patterns of behaviour and as a tool for stimulating the imagination.

However, a form of the latter becomes possible for the one who writes the narrative. Grant seems both fascinated by her mother's condition and at pains to make sense of it. She uses a series of literary metaphors to convey the otherworldly character of the dementia ward, which, despite Sontag's famous attack in *Illness as Metaphor*, become necessary for her in order to make her mother's condition more meaningful. When she and her sister Michelle first see the place, 'Michele looks round and instead of grandfather clocks, antiques and pot pourri, she sees death's waiting room'.[68] Unlike Swan's experience, this is mostly figurative. In other cases, Grant resorts to mythological allusions: '[My mother] wanted to be still in the world, the land of the living, but like Eurydice ... she belonged with the dying, abandoned in the twilight zone, and we could not go there with her even if we had wanted to'.[69] While these metaphors convey the daughters' guilt – 'What crime have we perpetrated, bringing her to this terrible place?'[70] – they also suggest that this space is heterotopian precisely in 'its sublimity, its transcendence of the quotidian'.[71] In Grant's words:

> That was the first visit when my mother stopped being my mother and became someone else, one of those individuals who are defined by the term Old Person. It wasn't just that she was in an institution but that the place itself was outside the world, in a hinterland between death and life.[72]

But Jewish Care is not just like any other dementia ward. Heterotopias often bring together different temporalities, and in a space such as Jewish Care where there are Holocaust survivors diagnosed with dementia, these temporalities often get mixed: '[t]here was a man in another of their homes who had to be carried screaming to the showers, mistaking his wash for a final march to the gas chamber'.[73] Moreover, as Marco Cenzatti notes, groups that live in heterotopias 're-code these other spaces with their own informal and often invisible meanings, rules, and times'.[74] Linda's mother, for instance, often confuses doctors with the immigrant authorities her family had to negotiate with when they arrived in Britain at the turn of the twentieth century.

While Jewish Care is characterized by heterochronies, it also encloses several spaces that produce strange juxtapositions; the cultural collision of inmates who are Jewish and staff who are black (from the Caribbean and Africa), for example, can be seen as analogous to Foucault's garden in 'Of Other Spaces': 'all the vegetation of the garden was supposed to come together in this space, in this sort of microcosm'.[75] This cultural collision adds to the heterogeneity and strangeness of the centre: '[a]t important times such as festivals, families can't make themselves available so on Friday evenings you can have this strange phenomenon of non-Jewish staff lighting the candles and trying to carry the residents through the experience that doesn't mean anything to them themselves'.[76]

If Grant seems to stress the dehumanizing effect of institutionalization on her mother (the memoir cites Erving Goffman's work *Asylums* (1961) on several occasions), her story offers another heterotopia as a brief space of resistance, namely the department store. The shopping mall features in many discussions of heterotopias as the postmodern heterotopia of illusion *par excellence*. Mall shopping is described as 'a kind of quasi-cinematic spectacle'[77] where people can both 'search – and construct – meaning'.[78] Whether it is a dangerous consumption centre or a site of escape, especially for discontented youths, 'the terror of time and space evaporates ... at the mall'.[79] In Grant's memoir, the department store in which mother and daughter often find refuge not only provides a space of bonding but also one where, by existing 'in the here and now', Rose can be her old self again:

> I find I can love my mother when we shop together, when we lose ourselves and the past and future in a department store – nothing that belongs to time of any significance except the rise or fall of the season's hemlines or its shades or the width of lapels or the colour of lipstick. So we shop together, outside time, mother and daughter united each in our purposeful quest to do what we have always done, and which to her goes on making sense.[80]

Conclusions

This essay has focused on real spaces, in the sense of spatial or material entities while also drawing attention to the lived experience of illness and to social/cultural factors that modulate the heterotopian qualities of a particular place. However, it is important to keep in mind that Foucault first used the term heterotopia in the preface to *The Order of Things* (1966) in relation to language rather than external space:

> *Heterotopias* are disturbing, probably because they secretly undermine language, because they make it impossible to name this *and* that, because they shatter or tangle common names, because they destroy 'syntax' in advance, and not only the syntax with which we construct sentences but also that less apparent syntax which causes words and things (next to and also opposite one another) to 'hold together'.[81]

Indeed, there is room for further exploration of textual or literary heterotopias in relation to illness narratives that pose challenges to the existing repertoire of life-writing conventions.[82] First-person Alzheimer's and other cognitive impairment accounts, as well as stories of terminal illness, push narrative to its limits, opening it to fragmentation, amorphousness, silences and other forms of generic and formal experimentation.

The spaces in the three texts I have discussed may appear to curtail agency as they transform a person into a patient, but it is important to stress that these stories present 'an ethic of freedom, realized in terms of aesthetic practice'.[83] If, as Kevin Hetherington writes, heterotopias 'are the sites ... in which humans experience their limits of their existence and are confronted by its sublime terror',[84] narrating one's experience of dwelling in the heterotopias of illness creates 'a screen' through which the terrors of illness can be contemplated with distance by both author (whether the patient or witness of another person's experience) and readers. In this sense, illness narratives bring together the material and metaphorical aspects of space which Foucault's work continually examines.

6 WRITING THE LITTORAL

Abdulrazak Gurnah

The Indian Ocean and everything that is in it has lost its charm for me.

Joseph Conrad, 'A Smile of Fortune'[1]

The heterotopia is capable of juxtaposing in a single real place several spaces, several sites that are in themselves incompatible.

Foucault, 'Of Other Spaces'[2]

Disembodied Locations

Joseph Conrad visited Mauritius in 1888, when he was a ship's master on his first command. He was detained in Port Louis for nearly two months because of loading difficulties and was forced into an unexpectedly longer acquaintance with the island than he had anticipated. Over twenty years after that visit, by now an established writer and recovering from the completion of *Under Western Eyes* (1910),[3] he received a letter from an admirer who had read one of his early Malaya stories. Conrad, always ready to oblige his readers with the kind of products they liked, ransacked his memory for more of the same and decided to write a story about Mauritius. The result was 'A Smile of Fortune'. The story opens with the captain-narrator's ship approaching Mauritius 'a fertile and beautiful island of the tropics,'[4] where it is going to pick up a cargo of sugar. As soon as the ship is safely in harbour and before the captain-narrator has finished dressing for the shore, a ship's chandler named Jacobus comes on board and attempts to manipulate the captain into a business arrangement. This connection with Jacobus is at the core of the story. At first the captain-narrator takes this to be the same Jacobus the ship's owners had recommended to him as one of the foremost merchants on the island, only this turns out not to be him but his discredited brother. When the captain-narrator later meets the intended Jacobus, he is appalled by his rudeness and is rude in return, and the encounter ends with the two men abusing and threatening each other. Later, the captain realizes that the quarrel was a mistake on his part, because the elder Jacobus holds a monopoly on the bags that are used to transport sugar, and the captain cannot embark his cargo as there is apparently a shortage of such bags. It is during this enforced stay on the island in search of bags that the story unfolds.

At its core is the story of chandler Jacobus's past passion and the events which led to his disrepute. A 'wandering circus' – a metaphor for sleaze and vagrancy – came to the island and Jacobus, who is described as slow-moving and calculating, in other words not an impetuous man, became 'suddenly infatuated with one of the lady-riders'.[5] When the circus left the island, he followed her, abandoning his wife and child, and his good name. The circus-woman treated him with contempt, both as a man and as a lover, and in the end Jacobus returned to Mauritius, disgraced and with a child which, at first, he could not be sure was his. Then the mother also returned to Mauritius – the lady-rider – a renegade without contrition. Her affairs had gone bad and she sought out the man that she knew she could impose upon.

Jacobus's daughter with the lady-rider is called Alice. The other daughter from the abandoned family is mentioned but not named, and is insignificant to the events except as a living symbol of Jacobus's disgrace. In some ways, Alice is an even more powerful symbol of this disgrace because she is the product of it, the unmarriagable and outcast emblem of Jacobus's flaunting of social laws of association. In addition to her mother's vulgar status as a promiscuous lady-rider in a circus, there is an allusion to her racial impurity, delivered by the captain-narrator in a sentence of condensed hatred and disdain. Here is his description of Alice's resemblance to her parents: 'Those long, Egyptian eyes, that low forehead of a stupid goddess, she had found in the sawdust of a circus; but all the rest of the face ... was Jacobus, fined down, more finished, more expressive'.[6] The combined images of resemblance to the mother signify race and 'birth' or class, and hint at Alice's otherness.[7] Nonetheless, the captain-narrator is unable to resist an attraction to Alice, and his friendship and increasing infatuation with her gains him notoriety too and contaminates him with her social and racial ambiguity. This is the emotional tension of the story, the captain-narrator's struggle with an ambivalent passion for Alice, the lady-rider's daughter.

This is Mauritius in the 1880's and the story points to the decadence and decay of the 'old French families'[8] with their rigid snobberies, but the portrayal is done in such strong terms that they seem simply a superannuated class, a doomed aristocracy, a cliché. We know from Conrad's biography that he was well-received into the planter families when he visited there in 1888, so much so that he proposed marriage to one of the two sisters in the family he came to know best, not realizing that she was already betrothed.[9] In his 'Author's Note' to the 1920 edition of *'Twixt Land and Sea*, Conrad wrote that 'A Smile of Fortune' was 'the most purely Indian Ocean story of the three [in the volume]',[10] yet the passing of twenty years has obviously changed his view of the hospitable people he had spent so much time with and one of whom he had considered for a wife.

In any case, despite many misgivings and social notoriety, the captain-narrator obsessively returns to Jacobus's house and to his unstable daughter Alice. His visits even draw the attention of 'the very niggers on the quays'[11] who turn to look after the reckless lover of Jacobus's daughter. The 'niggers' make few and brief appearances in the story, going about their servile business and turning to look after the tarnished captain-narrator. There is an earlier example of the story's casually racist register in the description of '[a] lanky, inky, light-yellow, mulatto youth, miserably long-necked and generally recalling a sick chicken'.[12] There is something familiar about him, the captain-narrator thinks, and then it dawns on him that he has the Jacobus features. He is the elder Jacobus's son with an unnamed black woman. The elder Jacobus has 'many of that sort', an acquaintance of the captain 'jovially' tells him, but he is also a 'highly respectable bachelor … there had never been any scandal in that connection. His life had been quite regular. It could cause no offence to anyone'.[13] The presence of the 'mulatto youth' in the story is, among other things, to demonstrate the hypocrisy of this decadent culture that will outcast the daughter of a circus lady-rider but will tolerate a 'mulatto son', albeit as a lowly clerk in his father's establishment. 'Mulatto' offspring are nothing to be ashamed of, they will know their place. The younger Jacobus's attempt to infiltrate his daughter into polite society, on the other hand, is a sinister matter, duplicitous and subversive of social order. The description of the 'mulatto son' is also to demonstrate that the result of such unions is a degraded specimen: he has 'an exotic complexion' and a 'slightness of build', his 'Jacobus strain, weakened, attenuated, diluted as it were in a bucket of water'.[14] This incident and its low-key racism precedes the brusque dismissal of 'the old French families' on the island which was mentioned earlier – 'The emptiness of their existence passes belief' – and the two outcomes – 'mulatto youth' and Alice – are clearly linked as a demonstration of their hypocrisy.[15]

The story's split focus on the captain-narrator's unexpected passion for Alice and the decadent milieu of the 'old French' planter society, has the effect of disembodying the location: Port Louis, Mauritius. The side-glances at the 'niggers' and the story's only mention of 'the imported coolie labourers on sugar estates',[16] positions this crowd on the periphery of the story's vision, a hubbub off-stage which does not even frighten Alice's old relative and chaperone, who is inclined to be panicky and for whom newspaper reports of crime and thuggery in European cities feel more menacing. By the 1870s the population of Mauritius was 70 per cent Indian and 20 per cent the offspring of African plantation slaves, yet Conrad's island seems to be inhabited by Europeans and one or two grotesque 'mullatos' and a handful of 'nigger' servants and a few 'coolie labourers' who are out of sight in the countryside. In addition to the occasions already referred to, there are seven other references to the non-European denizens of Port Louis.[17] All the references are inflected in some way, either by the word used to name them or by a diminishing adjective, but above all, this mis-shapenned little multitude seems to be a minority of strangers. Is this ironic, a demonstration of the narrator's blindness to his reality? There is perhaps a small clue to suggest this. In the opening of the story, the narrator describes the spectacle of the island and the sea as the ship approaches it, and in this lyrical mode, he laments that the sea is used for war and commerce, in other words, for imperial activities, but then he forgives himself in this way: 'But, living in a world more or less homicidal and desperately mercantile, it was plainly my duty to make the best of its opportunities.'[18] So we might see this as foreshadowing his partial vision of Port Louis and his stumbling affair with Alice, a way of obscuring unsettling sights from himself. Or is this perhaps how Conrad remembers his two months in Mauritius, socializing with the Europeans in Port Louis, becoming entangled in love affairs and catching glimpses of a servile host?[19] Perhaps also, it indicates a certain way of seeing, a strategy of self-consolidation through representation which is familiar in the colonial narrative. It is an instance of an 'other space' being superimposed onto a real space to create a new meaning, which is characteristic of what Foucault describes as heterotopia.

From out to sea, the coast of East Africa is the western extent of the Indian Ocean world. This is so in a geographical sense, but also both historically and in important dimensions of its cultural practices. European-imperial representations of this littoral ('pertaining to the shore') and the Indian Ocean cultural archipelago of which it is part, have often not succeeded in conveying its complexity and particularity. In one sense, in the denial of the complexity of the story's location, 'A Smile of Fortune' illustrates this, although it is by no means unique in that respect. Is this failure to do with how the cultural reality of this littoral challenges the observer and brings forth incomplete accounts, establishing a narrative convention for evoking this landscape, or does the blindness derive from European/imperial self-importance in vision? Is the impulse for this 'failure' a desire to produce another narrative more suitable to European-imperial purposes? In one of those informative passages which punctuate the rhapsodic fugue of her life on a highland farm, Karen Blixen has this to say about the coast of East Africa:

> The cold sensual Arabs came, contemptuous of death, with their minds, out of business time, on astronomy, algebra, and their harems. With them came their young illegitimate half-brothers the Somali – impetuous, quarrelsome, abstinent, and greedy, who made up their lack of birth by being zealous Mohammedans ... The Swaheli [sic] went along with them, slaves themselves and slave-hearted, cruel, obscene, thievish, full of good sense and jest, running to fat with age.[20]

To a certain extent, the effortlessly scornful ideas figured in the passage above are informed by what was at the time (1937) a still-continuing nineteenth century European discourse which constructed this eastern region of Africa as in the grip of a malaise inflicted on it by Arab and Indian mercantile activities. This construction was derived in part from the copious explorer narratives focussed on this region which appeared in mid-century as a result in the intense interest in locating the source of the Nile.[21] The most prolific of these was Richard Burton.[22] In his books about his Central African explorations, he wrote with an ambivalent sympathy for the Omani grandees he had to deal with in Zanzibar en route to the interior, and with less respect for the Indian merchants. But it was Livingstone's journals which left a more powerful testimony. David Livingstone spent his last tormented years wandering Central Africa in the company of Arab and Waswahili traders. In his journals Livingstone was tortured both by his illnesses and by the knowledge that he was being kept alive by slave traders, and his journals describe scenes of horror and violence that he witnessed, or that he thought he witnessed. The authority of the imperial eye-witness account is not beyond question, but in their own time, such one-sided testimonials as Livingstone's would have been received with little reservation. Livingstone was already famous, both as an explorer and as result of Stanley's dramatic search for him in the Central African wilderness, and the publication of his Journals after his death would have given them incontestable moral authority. In himself, Livingstone brought together the Victorian explorer in search of knowledge, the missionary evangelist and good doctor, and finally the martyr. His exploits and death, and that of others like him, coincided and prompted an intensified interest in this part of Africa, both as an arena for imperial expansion, as well as one for Victorian evangelism. This discourse, especially in its evangelical genre, advanced the narrative that the area was captured by the rapacious interest of Arabs, and to a lesser extent Indians, and that only European intervention could ameliorate this. The tenor of this narrative did not change through the colonial period, even if economic realism and political necessities of European colonialism required a degree of courtesy and negotiation with the displaced coastal elites.

The construction of coastal cultures as the equivalent of Arab Omani trading practices allowed the homogenization of its people in the Orientalist trope of the sensual and cruel Asiatics and their 'Swaheli' slaves that we saw figured in the Blixen quotation above.[23] For a European writer like Blixen, her account of the coast or coastal people was just another element in the narrative of self-consolidation: at once knowledgeable and superior, in command of the ignoble, blundering and essential imperial other. This controlling register is sustained throughout in her account of native people, for example in her repeated assertion that 'all natives' are either this or the other in her summaries of them. Because Blixen's attention on this native other is so sustained, her text's partial vision and its extreme violence reveal its instability as a crumbling text, unable to sustain its truth except by suppression and distortion and over-statement.

See, for instance, how Blixen announces firmly: 'The Swaheli tongue has had no written language until the white people took it upon themselves to make up one'.[24] Islam, of course, would have brought writing to the coast, and much of the western Indian Ocean littoral was Muslim by the early sixteenth century when Vasco Da Gama's Portuguese fleet made high-handed contact with several coastal towns. As is well known, it was from one of these towns, Malindi, that he found a pilot who led him to Calicut. The Portuguese were entirely ignorant of navigational conditions in the Indian Ocean and of the significance of the monsoon winds, and would have struggled without the pilot, though having got this far, they would certainly have found their own way there if they had to.[25] It is evident that Islam had taken root much earlier than this period, with some historians suggesting dates as early as the ninth century.[26] Blixen was not to know what modern historians now know, but to speak with such dismissive assurance about a matter she knew little about but could easily have checked when she was writing her account years later, confirms preconceived notions on the subject.

But there is also a hint of unexpected comedy in the moment that this remark on the 'Swaheli tongue' introduces. Blixen describes how she read letters aloud for her illiterate workers, and even though she could not speak 'Swaheli' she had no difficulty with this task because it was 'the white people' who had given it its orthography. She continues:

> with care it was spelled out as it is pronounced, and it has got no antiquated orthography to entrap a reader. I would then sit and read out their writings orthodoxly, word for word, with the receivers of the letters in breathless suspense all around me, and could follow the effect of the reading without in the least knowing what it was all about. Sometimes they would burst into tears at my words, or wring their hands, at other times they cried out with delight; the most common reaction to the lection was laughter, and they were continually doubled up by convulsions of laughter while I read.[27]

It is not clear that Blixen is aware of the possibility that her audience was laughing so much because of her delivery of the uncomplicated words. She follows up this moment with reflections on the mystic power of writing for her illiterate listeners, for whom the appearance of their names in writing has the quality of experiencing creation. The agent of this Adamic epiphany is Blixen herself, or at the least, European knowledge, which is here figured as writing.

The two examples I have been discussing illustrate a more general distortion and suppression in the representation of colonized locations and cultures, and many uglier examples than this are available. The intention is not to reiterate the falsifications of the imperial text in general. Conrad was writing at a time when European imperialism was at its peak, and he wrote in the register that was prevalent in his time. His achievement was to take the adventure novel of empire, with its mixed tales of European daring and abjectness, and bring to it a series of ethical questions about imperialism. By the time Karen Blixen was writing the memoir of her African farm, European imperialism had begun to lose the ethical conversation with itself some time before. It was not as easy as it had been to represent imperialism as progress, and impossible to reconcile Europe's self-construction as the champion of progress, while coercing other people and cultures in such blatantly unenlightened ways. My purpose in selecting these two examples, or rather to emphasize the dimension of these texts in the way I have done, is to demonstrate how the representation of the western Indian Ocean littoral and its archipelago in the imperial tradition, have imposed another reading of these locations, even in texts that are ambivalent in their sympathies.

A Displaced History

I have suggested that these texts are part of a convention of representation with its history in a discourse of contestation which was part of a process of imperial consolidation and legitimization. I will now turn to two nearly-contemporary texts which also address this region: the novel *A Bend in the River* (1979) by V. S. Naipaul and the travel book *North of South* (1978) by his younger brother Shiva Naipaul. My contention will be that, rather alarmingly, we find a continuation of some of these narrative flourishes in these texts by the two Trinidad Indian brothers. By an interesting coincidence, these texts appeared within a year of each other, even though the experiences on which they were based were more widely apart in time. *A Bend on the River* (1979), was in part based on V. S. Naipaul's experience of living in Uganda and travelling around the region in the late 1960's, as well as on a more recent visit to the Congo in 1975.[28] *North of South* describes and reflects on the events of Shiva Naipaul's travels in Kenya and Tanzania, and a fraught journey to Zambia, which he undertook a few months before the publication of the book.

The central figure and narrator of *A Bend in the River* is Salim, an Indian Muslim from the east coast of Africa. In the post-independence panic for Indian commercial interests in East and Central Africa, Salim buys a shop from another Indian, Nazruddin, who has decided to sell up and leave. The shop is in Kisangani, on the bend of the River Congo, the old Stanleyville and Conrad's 'Inner Station' in *Heart of Darkness*. In Conrad's metaphorical topography, the Congo signifies an unknown landscape as well as the hidden and residual space of human savagery. *A Bend in the River* makes reference to this topography, in part because Naipaul had at this point come to see Conrad as an important precursor in his reading of the colonial landscape, and what he saw as Conrad's sympathy for 'all men in these dark or remote places who ... are denied a clear vision of the world'.[29] Earlier in the same essay, Naipaul had remarked on Conrad's understanding of the colonized world: 'And I found that Conrad ... had been everywhere before me. Not as a man with a cause, but a man offering ... a vision of the world's half-made societies as places which continuously unmade themselves, where there was no goal'.[30] The idea of a people and a world without history appears repeatedly in Conrad's writing, most notably in the story 'Karain', which Naipaul discusses in the essay.[31] Naipaul's elaboration is to contrast this nihilism with the European 'habit of looking' and a capacity for reflection, an observation which is developed fully in the second chapter of *A Bend in the River*. As an example of this phenomenon of 'looking', he has Salim describe how the 'British Administration gave us beautiful stamps ... which depicted local scenes and local things'.[32] This, Salim says, is how Europeans with their detached vision, can transform a familiar scene by noticing it and remarking on it, as if saying: '"This is what is most striking about this place"'. The vague description of the 'postage stamps of our area' which the 'British Administration gave us', probably refers to a stamp series issued on Zanzibar Independence Day on 10 December 1963, one of which was of an outrigger in full sail skimming over the waves.[33] It is perhaps of no consequence to Naipaul that the person who painted many of these 'local scenes' was not the 'British Administration' but Maalim Abdalla Farhan, a Zanzibari artist and art teacher, renowned in his homeland for painting a whole oeuvre of such scenes, which in addition to the 'Arab Dhow', included the stamp series referred to above showing 'local' mosques, the Catholic and Anglican cathedrals and the Hindu temple.

Naipaul's point is to suggest that this is a culture that had allowed history to overtake it, and all the while it was unaware because it did not know how to 'assess' and reconsider its circumstances in the light of European domination. There may well be a case to be made for the failure of the coastal communities to understand how weakened they were at the point when colonial rule ended, but there were many reasons for this and it is not the case that the ability to 'assess' would have made much difference to the outcome. What is striking in Naipaul's rendering of their dilemma is the language of cultural and historical dispossession reminiscent of colonial discourses. It seems to me a comparable gesture to that executed by Karen Blixes when she denies 'the Waswaheli' a written language, to name but one textual parallel.

But if the Congo as location recalls Conrad's idea of colonized societies whose people are unable to have 'a clear vision of the world', an idea that Naipaul shares, the latter also saw the Congo as an archetype of another kind, that of the sinister and congenitally malfunctioning postcolonial state, a vision that had become dominant in contemporary western writing about Africa without his help. This way of thinking about the Congo is everywhere evident in the essay 'A New King for the Congo', where Mobutu is satirized for his bombastic egotism and the Congolese are the resigned participants of an unstoppable chaos. This chaotic location is where most of the events of the novel take place,[34] but the location and Salim's reading of it is inflected by his upbringing and reflections on the east coast. As we have observed, the Indian Muslim coastal milieu is described in the second chapter of the novel, most strikingly in the account of the extended home Salim grows up in a town very like Mombasa. These Indian Muslims are 'distinct from the Arabs and other Muslims of the coast'[35] and more like the Hindus.[36] As we have seen, Salim refers to them as a people who have no sense of history, to whom '[t]he past was simply the past'.[37] Salim extends this observation to the point of making the claim that 'history' with dates and a connected narrative is a European practice: 'All I know of our history and the history of the Indian Ocean I have got from books written by Europeans'.[38] This is not the place to correct Salim's (and Naipaul's) ignorance about textualizations of Indian Ocean history, but one of the dramatic revelations of Portuguese interest in the Cape route to India was how little they knew about the Indian Ocean when a great deal already existed in Arab and Persian texts.[39] It is not clear whether this lack of a historical sense refers to all the coast Indians or only to the Muslim ones, but the idea is a familiar one in Naipaul's discourse, both about India and about the Indian diaspora.[40] Here, though, it has a particular dimension of self-neglect at a moment of peril, the postcolonial moment of the African's ascendancy to control. Paul Theroux reported on V. S. Naipaul's irritation and annoyance when he witnessed the way Indian businessmen and shopkeepers allowed themselves to be intimidated in post-independence Uganda,[41] and in Salim's analysis, it is lack of self-knowledge and self-awareness which makes them vulnerable. It is Salim who observes: 'we never assessed ourselves'.[42]

One who did was Indra, 'a great brothel man on the sly',[43] like V.S. Naipaul himself. It was through Indra, a son of a wealthy Indian family, that Salim understood the insecurity of their position and decided to buy Nazruddin's shop in Kisangani and leave. Patrick French, Naipaul's biographer reports that the source for Naipaul's insight into the coastal Indian cultural miliue was probably the month he spent in Mombasa as a guest of Jagdish Sondhi, whom he met in England in 1957.[44] Sondhi's parents' house had a squash court, just as Indra's does, and overlooked the creek, which means it was in the old part of town where the wealthy Indian ancestral homes were. The month-long visit would no doubt have given Naipaul a good view of that community and its ways of thinking. French also reports that Naipaul became friends with James de Vere Allen, the eminent Swahili historian whom he met in Uganda.[45] De Vere Allen was a champion of the integrity and richness of Swahili culture, to the extent that his work down-played the influence of India and Arabia which is the accepted historical wisdom.[46] It would not seem as if de Vere Allen had much influence on Naipaul's reading of the Waswahili, who in *A Bend in the River* are characterized as slaves who wanted to stay slaves. The chief representative of this way of reading the Waswahili is, of course, Ali, the household servant-boy who re-names himself Mettie.

We meet Ali for the first time in the description of the family home I referred to earlier and which I now turn to briefly. It is a long passage, in which the argument evolves into dramatic summaries of miscegenation and power. The passage opens with a description of how dependent the slave remains on the master, even after the suppression of slavery – 'The slaves, or the people who might be considered slaves, wanted to remain as they were'.[47] The description is familiar as the infantilizing argument of planter revisionism as well as the psychologization of the resistance to decolonization, as for example in Mannoni's reading of Caliban's rebellion in *The Tempest*.[48] Naipaul's Salim has an intimate elaboration on this idea of dependency:

> [W]hen I hear the word 'slave', I think of the squalor of our family compound ... all those people, someone always shrieking, quantities of clothes hanging on the lines or spread out on the bleaching stones, the sour smell of those stones running into the smell of the latrine and the barred-off urinal corner, piles of dirty enamel and brass dishes on the washing stand in the middle of the yard, children running about everywhere, endless cooking in the blackened kitchen building. I think of a hubbub of women and children, of my sisters and their families, the servant women and their families, both sides apparently in constant competition; I think of quarrels in the family rooms, competitive quarrels in the servants' quarters. But they weren't ordinary servants, and there was no question of getting rid of them. We were stuck with them.[49]

It is remarkable how much that description is like Hanuman House, the Tulsi home in *A House for Mr Biswas*, until we reach the reference to servants. It is inevitable perhaps that there would have been a great deal that was familiar in this diasporic Indian family to a Trinidadian Indian. The observation about servants who prefer to be called slaves because that gives them greater status and makes them part of the family, takes the analysis in a different direction. In addition to giving us Ali as a slavish archetype: a childish and unpredictable fantasist, in fact, the excitable native of colonial discourse, Salim's reflections on slavery also allow him to make a comparison between Indians and Arabs:

> The people in our servant house were no longer pure African. It wasn't acknowledged by the family, but somewhere along the line, or at many places along the line, the blood of Asia had been added to those people ... This, though was a transferring of blood from the master to the slave. With the Arabs on our coast the process had worked the other way. The slaves had swamped the masters; the Arabian race of the master had virtually disappeared.[50]

Cut off from their sources, the Arabs took wives among the Africans and the wives took them over, and in time their children swamped them and will eventually kill them. To see this process as a swamping is a profound misunderstanding of the complexly hybrid community of the coast. Naipaul's vision of 'the Arabs' as a homogenous group does not understand the evolution of the Waswahili from the fragmented communities, including the Arabs, who are its constituents. As with Blixen's summary of the 'Swaheli' – 'slaves themselves and slave-hearted, cruel, obscene, thievish, full of good sense and jest, running to fat with age' – Naipaul's Waswahili are also slave at heart, petty and scheming, and with no vision of the world. The descriptions of Ali, 'the boy', contain many objectifying and summarizing gestures which diminish him and any rationale he might have for his actions. The details of the journey he makes in terror to join Salim in Kisangani, for example, are overtaken by the quip that 'he had made in reverse the journey which some of his ancestors had made a century or more before'.[51] This rhetorical deceit, implying an ancestry for the Waswahili from the deep interior of the continent, is another historical dispossession which disguises an unwillingness to understand the complex longevity of the coastal culture.

As I observed above, the bulk of the novel's events take place in Kisangani, and here at last Salim meets real Africans, and the prose descends to a new level of reification. 'Fetish' and 'magic' represent African civility, and human relations are summary and brutish: 'a man could knock on any woman's door and sleep with her'.[52] Ferdinand, a central figure in the novel's representation of 'the young modern African', is brash, empty-headed and self obsessed, 'a jumble of all kinds of junk'.[53] In explanation of the bizarre acts he performs, Naipaul has Salim say: 'Ferdinand's an African', confident that his best readers will understand the trope cluster being referred to. Ferdinand's mother, Zabeth, also lived 'a purely African life',[54] giving off a strange smell which signifies knowledge of 'magic' and disappearing behind an impenetrable tangle of bush and river in her return home. If Naipaul's Africans speak, whereas Conrad's did not, it is only to condemn themselves for their duplicity, egotism and spitefulness.

In Search of Diaspora

In the year before the publication of *A Bend in the River*, V.S. Naipaul's younger brother, Shiva Naipaul, published *North of South* (1978), an account of his recently-completed travels in Kenya, Tanzania and Zambia. In the 'Introduction' he added to the U.S. edition and which is included in the Penguin edition, Shiva Naipaul quotes from the letter he wrote to his British publisher, explaining the reasons for making the journey and writing the book:

> The book will arise, I hope, out of my own concerns – or, if you prefer, obsessions. What do terms like 'liberation', 'revolution', 'socialism' actually mean to the people – ie., the masses – who experience them?[55]

In reality, Shiva Naipaul already had the answers to these questions, as is evident from the dismissive and long-suffering accounts he gives of the meetings he arranges with journalists and activists in Dar es Salaam, and from many other comments that he makes about the politics of the places he visits. There are no encounters with 'the masses', and no doubt the phrase was used ironically in the first place, parodying the duplicitous earnestness of post-independence rhetoric. A more likely explanation is to be found later in the book when Shiva Naipaul describes how he went to report on the Ugandan Asian refugees quartered in the bleakness of Dartmoor in December 1972.[56] He was so moved by what he saw there – 'my security was shaken'[57] – that he began to reflect on his Indian family in Trinidad and to see the diasporic connection with the people he had been interviewing. After introducing this memory, he debates his Indianness in Trinidad: he could not speak an Indian language, had not gone through any Hindu rites, knew nothing of the religion or caste, and as he looks back on himself, he concludes that when 'I left Trinidad at eighteen I was nothing'.[58] In his telling, it is that encounter in Dartmoor that prompts these thoughts, and most likely forms the explanation for visiting that other place, perhaps to investigate diasporic echoes of Trinidad. He concludes these reflections with a visit he makes to the home of a Gujerati merchant and his family in Mombasa:

> 'You people who went out to the West Indies mixed up', an old Gujerati merchant I met in Mombasa said to me. 'Here we did not do that. We kept to ourselves. We held aloof.' He spoke with pride. His family had been resident in Mombasa for over a hundred years but they had remained of India. Africa had wrought no discernible changes in them. So it is with most East African 'Asians'. *That* had been their great strength; and their fatal weakness.[59]

The description of the family recalls the opening chapters of *A Bend in the River*, a closed and overcrowded Indian world, menaced by post-independence hostilities. Here Shiva Naipaul is tempted to figure that closedness as a strength his own upbringing denied him, even as he also sees it as something he had been saved from.

Other moments also recall the doomed-Indian motif of *A Bend in the River*. Shiva Naipaul meets an Asian friend Ashraf in Nairobi who says: 'One day we'll all have to pack up and leave',[60] who recalls Indra who says: 'We're washed up here, you know'.[61] These repeated evocations of Asian vulnerability construct an 'Africa' of unrelenting and essentialized otherness, which paradoxically is thwarted by the impenetrability of the Asian world that it then seeks to destroy. This paradox is dramatized in the Sunday promenade when 'it seems the entire Asian population of Mombasa gathers on the seafront'.[62] The gathering seems 'clannish' to Shiva Naipaul, and dangerous as it excludes and excites African resentment because it celebrates the impenetrability of the Asian world. The book's sympathy for the Asians is striking because it is in such contrast to its view of Africans.

From the start of his account, as he surveys other travellers at the airport, it is clear from his tone that Shiva Naipaul disdains 'Africans', that is, 'blacks', and perhaps he is frightened of them. It is also clear that he is more comfortable and at ease with 'expatriates' (which is used to signify non-local Europeans), even if at times they come in for mockery too. He notices an 'expatriate' family in the departure area and remarks: 'the mood of Africa ... was already upon them'[63] which is evident in their self-assured detachment. He comes across a delegation of Zairean 'financial advisers', a 'contingent of ... well-attired blacks' excited by their duty-free plunder. Where the 'expatriates' are cool and self-contained, the Zaireans are ridiculous and childish. As the chaotic Air Zaire flight takes to the air (the stewardess is described as 'steatopygous') Shiva Naipaul makes eye-contact with 'a bearded expatriate' who shrugged over the antics of the Zairean passengers.[64] It is not clear if the shrug is the 'expatriate's' invitation to Shiva Naipaul to be complicit with his superior gaze, or if Naipaul records it as a self-endorsement, showing that the 'bearded expatriate' had recognized him and shared with him the mutually superior gaze.

As he looks out of the window, Shiva Naipaul sees first 'the ordered fields of Europe, safe, serene, blessed' and then later his 'first sight of Africa' is a 'gray desolation' of 'featureless bush country'. His disengagement from 'Africa' rarely relents. Later in his narrative he quotes Karen Blixen on the 'Natives': 'Natives were Africa in flesh and blood', and the quotation is allowed to stand as if it is an uncomplicated descriptive statement of the 'Native's' rootedness. Shiva Naipaul then contrasts this rooted 'Native' to Meja Mwangi's 'African urban man' in Going *Down River Road*,[65] who is 'as bereft of "roots" or "identity" as any of his slave-descended American and West Indian brothers. He could be in New York; in Kingston, Jamaica; in Rio de Janeiro; in Soweto'.[66] Like Blixen, Shiva Naipaul does not want his African 'detribalised'. The narrative also quotes from Elspeth Huxley, another tireless chronicler of settler triumphs in the Kenyan highlands, as she describes an episode of the savage's encounter with civilization – how the locals mistake a kerosene lamp for a fragment of a star fallen to earth.[67] The description is quoted without comment, as if it too is from a reliable source and has no other function. He quotes Huxley's comparison of the Masai and the Kikuyu, and cites her authority on the caricature of Kikuyu gullibility.[68] This becomes something of a pattern in the narrative which is dotted with racist anecdotes which are referred to sceptically and then gleefully repeated.[69] By the time Shiva Naipaul gets to Dar es Salaam, the only person he can have an intelligent conversation with is the owner of the empty beach hotel where he is staying, who is an 'expatriate' German. Everyone else he encounters is either a wordy and obstructive bore or a half-wit, and all are ridiculous. Towards the end of his travels he gives up on 'the masses' altogether, declaring loftily: 'Had I not learned, after all this time, that nothing in Africa had meaning? That nothing could be taken seriously'.[70] The last part of his travels, a short journey to Zambia, is in the company of a group of American and British 'campers', and the traveller is no longer obsessed by burning questions or even curiosity. He is disdainful and bewildered by everything he sees.

But earlier, there had been a moment of ambivalence in the narrative before it crumbled to the familiar safety of disengagement. After his time in the highlands and before leaving Kenya for Tanzania, Shiva Naipaul spends time on the coast. He calls this section 'A Spell on the Coast', although of course all the Mombasa encounters were also on the coast. By 'coast' here he means Lamu, a small island on the north Kenya coast which is the heartland of Swahili culture. In the middle of the West Indian traveller's long description of the town as a ruin, a sentence about 'facial features' suddenly stands out:

> Facial features range through all the gradations from the African to the Arab. A maze of shadowed passageways leads off the main thoroughfare. These are barely wide enough to accommodate two people walking abreast. Sleepy-eyed men lounge on the steps of small, unremarkable mosques. Many of the older houses are in ruin. I stand in insect-humming bush and look at mounds of falling masonry, at crumbling archways, at rotting timbers, at the eroded remnants of fine moulding on walls. In some of the ruins betel plantations have been established. The stone-built town peters out into colonies of mud huts lining twisting, muddy lanes. Beyond are rubbish dumps, guarded by long-legged, protuberant-chested scavenger birds.[71]

This is his first morning here, and he has slipped into a familiar register, constructing everything he sees as ugly and meaningless. Again and again he returns to what seems to him a paradox of naming and 'facial features'. He meets 'a beach boy' who laments of persecution by 'the Kikuyu' government, which threatens to send people like him back to Arabia because they made black men into slaves. The beach-boy, Shiva Naipaul understands from this, is counting himself as an Arab, and in a one-sentence paragraph positioned to announce something of some importance, he says: 'I looked at him – woolly-haired, flat-nosed, thick-lipped: "Arab" here was a state of mind.'[72] A man is introduced to him as an educated guide, Ahmed, who befriends him and takes him to Manda island, a key archaeological site in researching the development of coastal Islamic culture. During the tour, a weary Shiva Naipaul remarks with metropolitan hauteur: 'Ahmed leads us from one pile of rubble to another.'[73] Ahmed is described as follows:

> He has a fine 'Arab' cast of features – prominent hooked nose, high cheekbones, firmly moulded mouth. Black-brown hair curls stiffly back from a broad forehead. His skin, smoky bronze in colour, seems to be irradiated with the sullen heat of the coast.[74]

Then suddenly Shiva Naipaul is unexpectedly sympathetic. Looking on with a tolerant eye, he says:

> Islam and Africa had coalesced, they had made something of each other which, however modest and provincial the scale, was authentic. Their contact had ended in

neither recoil nor parody ... The Swahilis were not imitation Arabs. They were Swahilis, participators in a great current of civilisation. They evolved an architecture, a poetry, a music, a style, that was quite unique ... They had succeeded in fashioning a personality of their own.[75]

Here is an example of a crumbling text: we have heard nothing of this Swahili personality but of 'the atrophy of creativity' and the crushing of the spirit under the weight of 'racial' resentment. Now the suggestion is of something enduring and 'unique'. The traveller is bewildered by the paradoxes that he cannot read but still insists on reading in the way he is familiar with. We see this in the description of the 'sikukuu' fair, the Idd celebrations. He can only describe ruins and tawdry entertainments in a town of 'blank facades with their intricately carved doors'[76] which remain closed to his gaze.

If Conrad after twenty years could only remember a handful of 'niggers' in Port Louis, and if V.S. Naipaul with his penetrating but jaundiced gaze could only see one homogenous crowd of 'Arabs' and the triumph of the West, Shiva Naipaul grasped that there was more to what he was seeing than he fully understood. His way of dealing with his lack of understanding was to fall back to the familiar clichés of imperial disdain with which to narrate the terrain.

7 HETEROTOPIA AND THE CRITICAL CUT

Iain Chambers

> People can only imagine themselves in empty homogeneous time; they do not live in it. Empty homogeneous time is the utopian time of capital. It linearly connects past, present and future, creating the possibility for all of those historicist imaginings of identity, nationhood, progress, and so on that Anderson, along with others, have made familiar to us. But empty homogeneous time is not located anywhere in real space – it is utopian. The real space of modern life consists of heterotopia.
>
> Partha Chatterjee[1]

These incisive words from the Indian historian Partha Chatterjee clearly evoke the voices of Walter Benjamin and Michel Foucault. They draw us into considering a modernity that is not simply doubled by subaltern actors and forces seeking to contest the hegemonic version of a history that insists on a unique telling. Recognizing the intersecting and planetary distribution of difference, location, and singularity there here emerges an understanding of the constellation of modernity that is disseminated in shifting rhythms along multiple scales and within the combinations of heterogeneous powers and practices. Against the empty dream of an utopic alternative promoting withdrawal from the seemingly unavoidable impositions of actuality, the instance of heterotopia proposes that we step out of an existing version of time to drop deeper into the folds of the contemporary world; there to assay and acquire its potentialities. Here, time is split from itself to permit the registration of other temporalities; an imposed and seemingly inevitable futurity is marked by the return of further, unacknowledged times. No longer the victim of a rigid archive, confined to the predictable rhythms of a numbing tradition, the past here becomes a vibrant t/issue that interpellates and interrupts the present. The authorized combination of materials fall apart, the archive is unlocked and its documents, voices, objects and silences scattered over altogether more contingent maps. Set to diverse rhythms and imperatives, the past comes to be configured by present urgencies in an emergent critical space. In this sense, the present is still emerging, still in the making: understandings have not yet docked; they are still under way, open to contestation, redirection and reformulation.

To step sideways, and remove oneself from the implacable logic of a single-minded modernity, is to step out of the cage of an abstract temporality. As a conscious cut, an alternative take, a blue note and deliberate dissonance, this idea of heterotopic thought and practice seeks to burrow below both the topographical logic of Foucault's disciplined spatialities and the eternal dialectic of narratives and counter-narratives. If we could consider remix as a method, just like jazz improvisations on the 'standards', the unhomely melody of the blues or the DJ's timely 'cut', then we can confront the sedimented and striated composition of a modernity that does not move to a single tune or uniform rhythm. In this form of historical and cultural mix we are encouraged to think more in terms of subjectivating forces, shifting combinations and unplanned vibrations, rather than remain locked in the power of established positions dividing the singularity of sense. The historical conjuncture is ultimately a performative space elaborated along multiple planes, diverse trajectories, and unpredictable depths. The *ratio* is neither linear nor transparent.

At this point, it is perhaps instructive to return to Foucault's noted radio talk on heterotopia and consider this statement:

> Heterotopias are most often linked to slices in time – which is to say that they open onto what might be termed, for the sake of symmetry, heterochronies. The heterotopia begins to function at full capacity when men arrive at a sort of absolute break with their traditional time.[2]

These are slices in time, intervals in the narratives, and interruptions in the machinery of truth. Foucault's arguments suggest far more than simply registering the question of spatiality and acknowledging the unregistered volume of the contemporary world. To cut space up into a heterogamous assemblage is also to multiply time in diverse rhythms and temporal insistences. Neither is homogeneous. The desire to represent modernity as the perfect match of linear time and homogeneous space is thwarted. This teaches us that critical labour is not simply about contesting the imposed temporalities of hegemonic rationalities. Rather, critical labour is also about constructing a space besides us; here the reconfiguration of actualities releases another set of spatio-temporal coordinates, and another manner of reasoning. What now returns to the map is what was excluded by the premises of the previous cartography of power, of knowledge. This operation, like a Deleuzian 'fold', creases and deepens spatiality while rendering diverse temporalities proximate. It produces the interleaving of heterogeneous dimensions that resonate with the circulation of bodies, histories, cultures and capital in what we come to understand as a manifold modernity.

In this sense, as Foucault insisted in the Preface to The *Order of Things* (1966), the heterotopic breaks through and beyond the homogeneous order of a discourse.³ The heterotopic proposes the elision of the imposed. Language and space no longer match: the former is unable to contain and control the latter. The map is no longer reality, merely a limited representation. It is at this point that the epistemological device – the map, the discipline, institutional knowledge – promotes an ontological rift. For if utopias are the product of language, heterotopias are manufactured and maintained in the mutable materialities of space where language seeks to impose its order. If utopias by their very nature do not exist, they are nevertheless cultivated and cared for in language. Heterotopias, on the contrary, even if unregistered and unrecognized, do exist. They, too, require language and are therefore not without their utopic drive, but they sustain diverse experiences of inhabiting history and culture, different practices of time and space. Alternative, subaltern, and subordinated to the rules that occlude their presence, heterotopias exist and persist as counter-spaces beside and outside the dominant syntax of sense. Their presence uproots the premises of the linguistic and discursive order, proposing a flight and a freedom from its imposition. They propose a disturbance, an intimation of the unhomely. If space is produced (Lefebvre) and never simply given, it is not only produced by our language.⁴ It is construed, constructed, crossed and signified by many different bodies (and not all of them human). This hetero-genetic understanding of the spaces and temporalities of modernity is clearly irreducible to the mirror of Occidental power and conceit. The actors and agency involved are not merely of European provenance. The 'world picture' (Heidegger), proposed by European humanism and secured in the historical expansion and cultural domination of colonialism, imperialism and Occidental capitalism, no longer provides a seamless fit between representation and reality. What once lay outside the frame has entered the picture, traverses and troubles the perspective. The field of vision turns out to be an agonistic space in which borders and belongings are increasingly fluid, tactical and transitory; hence the increased application of violence – both physical and juridical – to control and direct them.

No doubt such comments stretch and even desert Foucault's original discussion of heterotopia by pushing experiments of the self and the social out into the exposed spaces and temporalities of the postcolonial world. Still, this return to Foucault in an altogether more extensive landscape also allows us to pick up further items from his *oeuvre* that clearly continue to vibrate in contemporary critical circuits. Here, confronting the 'global colonial archive'⁵ produced by Occidental culture and capitalism, Foucault's attention to the productive relations of power, to differentiated bodies and their governmentality, provide essential critical levers for re-opening that vault. When power is translated into the power to exploit subordinated bodies and extract wealth by conquest and dispossession, then the archaeology of 'progress', and the unilateral modernity of the European museum and its accompanying anthropologies, sociologies and histories, are exposed as provincial inventions. These local knowledges disavow the historical, cultural and geo-political coordinates of their production. Invested with a universal truth that disciplines the other and the elsewhere (and the mutual contamination of religious and secular conviction is here extremely pertinent) transforms a provincial and problematic practice into an unaccountable fetish.[6] What is represented is also what is repressed in the arbitrary violence of the unilateral, the unique, the universal. The rest of the world has been forcibly conscripted to that modernity, and there consistently subordinated in order to be exploited, catalogued, cast overboard, even exterminated; everywhere reduced to the deadly calculus of objectified value in the abstract circuits of capital. This is not a moral judgement, but a structural and historical one. It proposes not simply an abstract ethical horizon, but also, and more insistently, a living political one.

Maps, as Sandro Mezzadra and Brett Neilson have most effectively pointed out, are epistemological devices of profound ontological significance.[7] In their modern history, which is not incidentally the history of European expansion on a planetary scale, maps are pictures of power, representations of rule. They establish not simply territory and property, but also cognitive borders that seek to locate and control bodies and cultures both at home and abroad. Or rather, given that such rigid spatial distinctions are increasingly undone, they constantly trace and indicate the flows and fluxes of tangible and intangible mobilities on multiple scales in planetary frames. In this history, the history of modern Occidental cartography, power is masked in maps where murder is presented as measurement, and genocide become geometry. Maps, as such, are the transposition of the violence of the commodity form into the implacable 'laws' of the world market.

All of this means, as Chatterjee insists, to break with historicism. It is to break with a manner of narrating time and space as though from a unique point of view: Europe, the West and its accompanying configuration of knowledge (historical, sociological, anthropological ... economical, in sum, the assemblage of the human and social sciences). In a planetary political economy of knowledge, those 'sciences' that have largely disciplined and explained the world are not cancelled but rather re-worked in response to an altogether vaster cartography. The methods that apparently guaranteed their universality, once re-located in the specific historical and cultural formation of modern Europe since 1500, are exposed to interruption and interrogation by 'foreign' forces. Borders are contested and crossed, disciplines unwound, language retuned. Of course, the power of this inherited assemblage is formidable. Practised and institutionalized on an increasing world scale, over centuries of military, political, economical and cultural domination, this mode of understanding and its claims to universal validity remains hegemonic. Present-day academic labour, refereed journals, national research assessments and global university rankings all testify to the ubiquitous 'neutrality' of this exercise of power. Yet it is ultimately a deadly mechanism, able only to rationalize once it has 'killed' and objectified the body, the event.

It analyses in a manner that reduces the complexity of life to the confines of the discipline, transforming the opacity it seeks to explain into a moribund transparency, shunning vital ambiguities for the still life of an abstract verdict. As a knowledge formation it privileges the protocols of its reproduction over engagement with what precedes and exceeds its ken. Ultimately, despite detailed description and verifications, it opts for the comfort of conclusion rather than the disquiet of a critical opening. Knowledge is reduced to the procedures and protocols of the method. What are sought are a homecoming rather than the journey, self-confirmation and not the exposure of the encounter.

What, then, are these other forms of knowledge? Where and how are we to locate these unrecognized epistemologies of modernity? Insisting on the idea of the heterotopic, is to insist on understandings, knowledge formations and practices that may well be marginalized and unregistered and yet exist and resist besides us as part of the modernity we consider our own. If we were to employ a Freudian topography this would constitute the sedimented layers of modernity's unconscious. Perhaps, however a more horizontal topology better illustrates the unrecognized heterotopic proximities within modernity. The critical knowledges that emerge in the unrecognized dynamics and mutations of a planetary modernity can no longer be considered simply the provenance of the psychically repressed and physically distant. The dark depths of what was once suppressed in the peripheries of the world can no longer be relegated to the posterior conclusions of the discarded colonial map. Today, administrative and geographical information – European governmentality – is unable to block other forms of knowledge entering the picture and disturbing the narrative frame. Of course, the world continues to be ruled from that particular house of knowledge, but it is hardly a truly critical knowledge, rather more an administrative hubris that seeks to reproduce itself and, as a consequence, the status quo.

A modernity that is diversified, multiplied, and does not follow a single imperative, is a modernity that spills out of the homogeneous time of capital and the nation. Lived, that time acquires flesh and follows different directions. Its sense is neither merely inherited nor imposed. It is transformed into the urgencies of a body, a life, and a location. It is modernity whose transit reveals the centrality of translation, and is never simply mine to define.

Maritime Criticism

> Insofar as the academic discourse of history – that is, 'history' as a discourse produced at the institutional site of the university – is concerned, 'Europe' remains the sovereign, theoretical subject of all histories, including the ones we call 'Indian', 'Chinese', 'Kenyan', and so on. There is a peculiar way in which all these other histories tend to become variations on a master narrative that could be called 'the history of Europe.[8]

In his conclusion to his brief comments on heterotopia, Foucault famously nominates the ship as one of its privileged sites. The ship, the sea, together with a whole archive of Caribbean poetics (Glissant, Walcott), alerts us to a decisively alternative manner of crossing and configuring modernity. For attempts to politically and juridically 'fix' and frame the sea according to the requirements of terrestrial and territorial coordinates always goes adrift. It is the fruitful and suggestive critical nature of this 'drifting' and its floating premises that most clearly invest us with a series of questions that are perhaps rarely posed. Here, and most obviously, the status of borders and confines are continually challenged by marine transit, migration (both legal and illegal), encounters with opacity, and an emerging sense of belonging and an extended citizenship that is irreducible to the terms of the existing polity.

Commencing from the sea, rather than the habitual location of land and territory, is clearly to propose a de-stabilizing style of argument in which unknown factors, critical uncertainty and historical anxieties are deliberately foregrounded. Such a choice of perspective has much to do with deliberately seeking to unsettle many of the disciplinary procedures and protocols of the social and human sciences. Opposed to dreams of systematic order and the assurance of canonical convictions to be housed in libraries, museums, data-bases and syllabuses, what we might call 'maritime criticism' sets existing knowledge afloat: not to drown or cancel it, but rather to expose it to unsuspected questions and unauthorized interruptions. The presumed stability of the historical archive, together with its associated 'facts', and cultural identifications, invariably sealed in the narration of the nation, can be set to float: susceptible to drift, unplanned contacts, even shipwreck. As we have learnt from the rebellious histories of the modern Black Atlantic (C. L. R. James, Paul Gilroy), and the counter-histories suspended in its depths, in the sea are histories and cultures that are connected, rather than divided, by water. In that 'grey vault, the sea'[9] there lie other histories, other ways of narrating both a local and planetary modernity. Rather than refuse to engage with hydrophasia, with what the artist Allan Sekula has called 'forgetting the sea',[10] we could choose to return to the centrality of this liquid domain.

This is not merely a metaphorical whim. The sea has been consistently central to the making of Occidental modernity. Voyages of so-called discovery, European colonization and the global nets of imperialism have all been about the sea: mapping it, crossing it to colonize other worlds, controlling it to conquer global hegemony, travelling its sea-lanes in the pursuit of the reorganization of planetary labour (from yesterday's slave ship to today's container vessel), harvesting its resources (from fish and whales to oil and gas). Rarely, however, has it been considered in its own right as an ontological challenge to the histories and events that apparently require terrestrial ground in order to be narrated. Still the politics and poetics of the Caribbean, with its archipelago of creolizing histories passing from island to island, and then along subterranean circuits sustained in the signs and sounds of the bass cultures of the Black Atlantic diaspora, suggest an altogether different telling. If the modern Atlantic world is generally presented as the fulcrum of Occidental modernity – its political economy and institutions of democracy – it was simultaneously also the world of the slave trade and the proto-industrialized organization of plantation labour. Many of the Founding Fathers of the Constitution of the United States were not only slave owners (Washington, Jefferson, Madison) but also some were defenders of the cultural and legal legitimacy of the institution that transformed specific human beings into the fetishized anonymity of commodities (directly inscribed in several articles of the 1787 Constitution). If events in Paris in 1789 no doubt looked to the successful revolt against the British Crown in North America, that same revolutionary spirit was simultaneously loath to permit its universal message of *Liberté, Égalité, Fraternité* be adopted by black slaves in revolt in its richest colony in the Caribbean: Saint-Domingue, subsequently the black republic of Haiti. These are histories sustained by water and the global reach of the oceanic traffic in bodies, institutions and ideas.

The intensity of historical negation and the political energies deployed in the repression of such histories creates a black hole in the very heart of the progress that modernity supposedly embodies. Through the mechanisms of racialized discrimination claims of democracy fall apart in the refusal to permit its exercise in the lives and labour of others. Only a teleological acceptance of historical progress can be confident that this earlier state of affairs has finally been overcome. An understanding attentive to the genealogies of contemporary structures of exploitative power in Occidental modernity would be altogether less sanguine: Saint Domingue in 1791, South Africa in 1961 and Ferguson, Missouri in 2014 move, along with the rest of the Western world, in a shared matrix. We are not so much confronted with a moral choice but rather with the sedimented, structural articulation of power, privilege and its perpetuation. Haiti disappears into this black hole (politically quarantined and historically cancelled by the West throughout much of its history), allowed only to appear in the alien alterity of abject poverty and vodou. Meanwhile, 'unauthorized' appropriations of the political and cultural language of modernity, orbiting around questions of freedom, liberty and universal rights, disseminate critical questions that connects the Eighteenth century Atlantic world to the modern Mediterranean one. Both slaves in Caribbean revolt and rebellion and modern migrants seeking to enter 'Fortress Europe' tend to be considered illegitimate intruders in the narrative. It is precisely here that the maritime archive sustains another accounting of modernity, casting a critical wave into the contemporary Occidental mechanisms of controlling and defining the Mediterranean.

French and British sea-borne empires fighting for global hegemony simultaneously in the Caribbean and the Mediterranean propose a colonial archive that continues to haunt the present. Nelson in the Bay of Naples crushes the Neapolitan Republic (inspired by the French Revolution and hence sharing unexplored coordinates with the slave rebellion in Saint-Domingue). The previous year he was in Egyptian waters at Abu Qir Bay (the so-called 'Battle of the Nile') destroying the French fleet that had transported Napoleon and his troops to Egypt to seize this Ottoman province (thereby establishing the European coordination of the Middle East, the question of Palestine and the eventual constitution of Israel). The historical passage from land to sea, and eventually air, power is not rehearsed here in order to confirm Carl Schmitt's resigned prediction of the end of European global hegemony.[11] Rather, between the solidity of territorial confines and the seeming immateriality of the atmosphere, the shifting surfaces, islands, continents, currents and depths of the marine world offer a critical interruption, an exit from the linearity of the technological conquest of space and a historical narrative that unfolds in the empty abstractions of universal time.

If today's Mediterranean has clearly become the marine cemetery of modern day migrant labour, it also opens up a maritime archive that draws us back to the Caribbean and its centrality to the modernity we think we know so well. From the Mediterranean to the Caribbean: from the histories of these waters emerge the testimonies of a political economy that reduces racialized bodies to the fetishized universality of commodities. Beneath the waves, on the other side of the official chart, are the anonymous processes and peoples making the modern world from below, from 'way, way below'.[12] These clandestine histories sign the unsuspected register of a heterotopic modernity.

So, we are invited to step off shore for a moment and consider inherited referents from elsewhere, in another time-space configuration and through other eyes; for example, that of the desperate vision of a modern-day migrant riding the waves in a small boat, adrift and gazing hopefully northwards.[13] Here the migrant's time becomes the migratory time of modernity. The distant shore and the marginal world that is hidden and ignored become immediate; it is literally figured and exposed by the body of the feared foreigner, the despised stranger. Like a nemesis, the interrogative presence of the migrant arriving from the sea announces planetary processes that are not merely *ours* to manage and define. He or she draws Europe and the West to the threshold of a modernity that exceeds itself. The migrant's time creates a slash in my time through which modernity itself migrates and subsequently returns bearing other histories, other modalities of narrating the modern world. Recently this rich complexity and accompanying ambiguity has been most beautifully caught in Andrea Segre's film *Shun Li and the Poet*(2011). The friendship that develops in a sea-front bar in the Venetian fishing port of Chioggia between the Chinese Li and the older fisherman Bepi from ex-Yugoslavia throws an interrogative shadow over the belligerent localism of the lagoon. Understandings of belonging are exposed to other winds and currents; home is constructed on an altogether more unstable location; like Bepi's house, perched on stilts on the choppy waters of the bay. Left to Li after his death, she burns it down: a flame temporarily flaring up over a watery world.

In the fluidity of such spaces the migrant emerges not merely as the historical symptom of a mobile modernity; rather *she is the persistent and condensed interrogation of the true identity of today's planetary political subject*. At the end of the day, his or her precariousness is also ours, also *mine*; for it exposes the coordinates of a worldly condition in both the dramatic immediacy of everyday life and in the arbitrary violence sustained in the abstract reach of the polity, the law, and their provincial entrenchment. Citizenship is not a permanent state; it is a precarious and localized one. Ultimately, it is the contemporary migrant, and her clandestine histories, who re-opens the archive and ushers repressed histories into the present. While drawn back we are simultaneously sent spiralling sideways beyond the screens of progress into Europe's violent heart of darkness.

The Art of Heterotopia

> since Columbus, Europe had been obsessively engaged in voyages of self-discovery requiring it to try and match the coordinates of intercontinental space by those of universal time – geography by history.[14]

The fluid archive provoked by maritime criticism, and the maps of modernity traced by the modern migrant, suggest a historiography of not how things actually were, as though forever fixed in time and secured in a scientific 'objectivity', but rather of how things are, exist, survive and live on. Here there are artefacts, documents, material traces, but there are no historical 'facts' isolated from the human and social activity of appropriation and interpretation. As Walter Benjamin would have insisted, history is always now. The present is haunted and interrogated by the past. The dead continue to speak in the insistence of images – visual, scriptural, sonorial – loaded with time. A linear historicism that believes the past is really past, reduced to the dumb witness of an unchallenged and conclusive explanation of the present, snaps and unravels. This is to work with the idea of an interrupted and interrupting history, a history of intervals and discontinuities, of multiple temporalities. It leads into a shifting geography of memory (and forgetting) where meaningful details are connected with forgotten futures: a dynamic interweaving of past, present and future collated in the intensities of the present.

Out of the traces of time other configurations emerge to promote a diverse narration of both yesterday and today. Such memories of the futures, as we know after Freud, and so evocatively traced by Chris Marker in his film *Sans Soleil* (1983), are indivisible from the media that record them. They propose other historical records. The prevalent historical discourse is crossed by the insistence that history, precisely because it is always pertinent, always now, is not simply a matter that can be left behind to echo ineffectively in the institutional archive of the historians. What history and whose history is irreducible to a neutral catalogue of the past. The authorized tale is always lined with other memories: personal, collective, incomplete, and imagined. These memories, whether recalled or ultimately consigned to oblivion, are also where illegal passages and clandestine tellings seek alternative narratives of belonging and becoming. As Ranjit Guha has so effectively pointed out, this is to recover and subsequently subvert the Hegelian concept of the 'prose of the world' by insisting on its right to query the philosophical privileges of the 'prose of history' embodied in the modern European nation state.[15] The transformation of the negative identification of a 'people without history'[16] into a positive critical injunction is precisely where the extra-European world becomes an interrogative and historical force *within* a modernity that presumed its necessary exteriority.

In other words, alongside a supposedly factual economy, and its conclusive empiricism, is an *affective* one that recognizes itself in inconclusive interpretations, and which seeks its rigour in the complexity of the historical locality and cultural constellation in which it moves. It is precisely here that the artwork, for example, can be considered not merely as testimony to a historical past and present, but rather provides and provokes a diverse configuration of time, being and becoming. We are not dealing with ornaments to add to the nitty-gritty and cruel drama of historical and cultural formations. Details and fragments – shafts of condensed time – sustained in artistic languages, in a poetics, propose, as both Walter Benjamin and Aby Warburg argued in their different ways, another and radically different way of understanding and interpreting such formations. This is what the artist and critic Bracha Ettinger calls *artworking*.[17] Music, the visual arts, poetry, literature, even recipes and food, are neither merely metaphorical nor symbolical. In their material affects and historical insistence, such languages, through their excessive and unauthorized reach, become critical. They are able to disturb and displace the authority of the disciplinary accounts provided by historiography, sociology, anthropology, art history and literary criticism. At this point, the image – whether visual, textual or sonorial – is less the object and attention of thought and more the instigator of thinking. From the detail, the fragment of condensed time, from the dynamics of an image, the dissemination of sound, it becomes possible to rethink a space: the Mediterranean, modernity, Europe, the contemporary world. This would be to propose an unsuspected sense of place and belonging that can no longer be presumed to be of a unique origin or singular explanation. Against the desire for conclusive transparency, there is an excess of language, which in its undisciplined reach promotes a poetics and a further, unsuspected, politics. Such an interval opens on to a critical space that runs through, alongside and beyond the immediate pragmatics of a regimented time and place.

The veracity of the image is now to be located elsewhere, it is no longer a simple support – realism, mimesis – for narration, but is rather itself the narrating force. There are not images *of* life, but images *as* life; a life already imagined, activated and sustained in the image. There is not first the thought and then the image. The image itself is a modality of thinking. It does not represent, but rather proposes, thought. This is the potential dynamite that resides within the image: it both marks and explodes time. This is the unhomely insistence of the artwork working a critical cut, inducing an interruption. In the artwork, in the movement and migration of language, denomination is sundered from domination as it unfolds along unsuspected critical paths through the folds of a de-possessed modernity.

Musics and Maps of Heterotopia

Between the Black Atlantic and Mediterranean blues we can chart an ecology of rhythms, beats and tonalities that produces sonic cartographies where, as Kodwo Eshun and Steve Goodman have put it, 'sound comes to the rescue of thought'.[18] Drawn from a blue archive that plays and replays modernity, exploiting the spaces between its official notes, unsuspected sounds and sentiments cross, contaminate and creolize the landscape: in the Caribbean, in the Mediterranean ... in the modern world. Such musical maps provoke forms of interference that render hidden histories and negated genealogies audible, sounding them out and rendering them sensible. Not only do sounds matter, they also propose and extend critical matters. Sounds become a narrative force that draw us towards what survives and lives on as a cultural and historical resource able to resist, disturb, interrogate and fracture the presumed 'unity' of the present.

To think of the Mediterranean in terms of its sonorial suspension, and the unsuspected deepening and dispersal of the empirical present, is to embrace what Gilles Deleuze and Félix Guattari would have called a 'minor' history.[19] Sonorial cartographies and musical maps provide us with an interruption, or slash, in existing cartographies. Returning us to what has been overlooked, such musical maps permit another story and an unsuspected landscape to emerge. If established powers refuse to listen, as they inevitably do, then these sounds trace another, largely, unrecognized and undisciplined projection that shadows and potentially interrupts the seamless surface of consensual understanding. The affects of the artwork push the premises of historical and sociological analysis beyond their explanatory frames. What fails to be figured in such disciplinary terms nevertheless exists and persists as an interrogation, a potential interruption. Here, considering the Mediterranean as an 'infinity of traces without ... an inventory',[20] sonic histories propose a persistent 'noise' that disrupts the institutional silence of the historical register. Sounds become a source of critical disturbance, and the musical archive 'a question of the future itself, the question of a response, of a promise and of a responsibility, for tomorrow'.[21]

Like the sea, that once facilitated their passage, sonic processes resist representation and propose an affective economy, 'stripped of consolation and security'.[22] They are inherently diasporic, destined to disturb fixed configurations of time, space and belonging while involved in sounding out communities to come.

Against a stereotypical image, Mediterranean sounds, as they travel, disseminate and differentiate, propose a complex cultural and historical place that can be heard in a diverse and altogether more open manner. This, for example, is what the great Palestinian critic Edward Said realized in his adult rediscovery of the voice of Umm Kalthūm.[23] Trained in the well-tempered aesthetics and classical music of the modern West, the lengthy improvised vocals of the Egyptian singer initially seemed to him to propose a sound on the edges of critical nonsense. Subsequently repositioning himself in and to the West, and hence to the voice of the excluded, Said drew from this unsettling experience the suggestive perspective of a multiple and *contrapuntal* modernity.

To respond to the voice of Umm Kalthūm is not merely to recover a sound from the negated archive of Mediterranean musical memories, it is also to propose another critical compass. Her voice and music were not simply that of a particular Arabic musical and poetical tradition. The manner of her singing and the execution of her music were also profoundly modern. If her voice, with its extended musical lines, shifting intonations, *melisma* and the privileging of the performative, obviously draws upon a long history of improvised musical execution in the Arab world and elsewhere (European classical music is here the exception rather than the rule), it also resonates profoundly with other modern, improvised metropolitan sounds of the twentieth century: above all, with the musics of the black diaspora, with the blues, jazz and subsequent derivatives in reggae, hip hop, rap and urban bass cultures. Although largely unknown and unheard in the West, the music of modern, metropolitan Cairo between the 1940s and the 1960s is the sound of Umm Kalthūm. It is not the 'traditional' sound of a folk music, antecedent to a subsequent entrance into modernity. Umm Kalthūm was musically and culturally very much an innovative figure and a modern woman.

Regularly transmitted on the radio, ceaselessly recorded, and presented in innumerable public concerts, she was a popular and commercial 'star' and a persistent public presence in the Muslim world. Umm Kalthūm's musical existence in the modern Mediterranean suggests an unsuspected proximity with other sounds, other places, other histories, that the altogether more strictly confined understandings of musical and cultural identification has structurally avoided and negated. This Mediterranean, as a complex cultural and historical formation, presents us with a 'unity in difference', where, as the musicologist Bruno Nettl notes, the challenge of heterogeneity is nevertheless characterized by certain common traits: for example, the ubiquity of plucked instruments (from the 'oud, and the guitar to the mandolin and the bouzouki'), or the strong imprint of Muslim music making with its modal systems and monophonic structures.[24]

Gilles Léothaud and Bernard Lortat-Jacob have pointed to the musical centrality of a fluid East-West axis that marginalizes the rigidity of the classical north-south division between a modern, Christian, Europe and a so-called underdeveloped, Islamic world.[25] They argue that the Mediterranean 'voice', in the hint of a cry (think of Arabic song, flamenco, Neapolitan vocals and rebetika), in its nasal intonation, in its dark, rough and granular textures, in the insistence of *melisma* rather than a distinct punctuation, the voice, on the edge, close to break-down, registers the exertion of the body in the song, and resonates in the ear as a distinct Mediterranean musicality. Opposed to the rigidity of many an institutional explanation and its static insistence on the securities of local and national myth, the fluidity of sounds propose an altogether more frayed and fluctuating map. This allows us to consider how the multiple histories of the Mediterranean are suspended and sustained in sounds. As instances in time they render its complex configuration momentarily audible.

Here the question of 'art' is no longer secured and explained in terms of aesthetic autonomy, historical mimesis and cultural representation, but, rather, promotes a critical instance in terms of sonorial power and the intensities of the affective economy it promotes. In the performative instance music proposes an acoustic or aural knowledge that restores the body to its multiple senses. A recovery of sensory geographies provokes a consideration of music in terms of a sensuous epistemology that touches on another, subaltern and suppressed, knowledge, carried in the body, sustained in sound, registered in rhythm, broadcast in the persistence of a 'bass history' (Linton Kwesi Johnson).[26] We are here engaging with an altogether more diffuse and less instrumental epistemology than that associated with the objectifying procedures of sight and the accompanying fetish of truth as representation. Music becomes less the object of thought and more the instigator of thinking.

The renouncement of reducing music to an object of historical, sociological and cultural attention brings us to reflect on the production of bodies through sounds. We are not here considering subjects that encounter sounds (subjects dealing with objects), but rather considering how the materiality of sounds are affectively sustained in individual and collective bodies via particular situations and milieus. Music at this point does not 'represent' a pre-existing state but rather promotes a sensual and social becoming. Sounds and musical practices alert us to a potential and irreversible breakdown in representation: the sound floats free and disseminates an unsuspected critical challenge. A particular modality of singing or musical genre does not simply represent or signify the cultural concerns and historical insistence of a precise community, social group or subculture. It does not simply do that. The sound is dispossessed. If music inevitably contributes to the cultural and historical tuning of the world it traverses, the fluidity and the differentiated immanence of its passage is also irreducible to a single moment of cultural authorization. In this sense, the sociology of music is displaced by music as sociology; the histories of sounds are replaced by the sounds of histories: sounds do not so much illustrate as propose histories. Sound narrates and affects an attachment to a memory, a place, a trace. It elaborates a temporary territory and a transitory abode in the world.

This, as Julian Henriques has suggested, leads to shaking 'the monopoly of rationality and representation'.[27] To consider music as proposing the potential of unsuspected territories of sense, secured in what is felt rather than fully explained, is also to challenge ideas of identities that seemingly root and resume the sound. In the prospect of a concentrated affectivity of materialities, sound, performance, reception and repetition tend towards a new, more vulnerable, more open critical instance. To reason in the archive of repetition, brings us into proximity with the concept in Arab music of the *tarab*: the ecstasy or enrapture that results from the stretching of vibrations through a reiteration that is never the same, that continually unfolds in becoming the song and the sound. Opposed to the isolated, individual achievement of abstract musical perfection, the performative possibilities of improvisation include the audience as it strives towards the creation of community. In the repetition, what Deleuze and Guattari nominate as the refrain, lies the temporary autonomy of a territory sustained in a sonorial event where music promotes a reasoning medium. Here where 'time is split from time',[28] music does not 'mirror' or 'communicate' a seemingly separate and independent reality, it is a *ratio*, a form of reasoning, rather than a representation. This inaugurates a history of the discontinuous destined to disturb presumed linearities and the assumed teleology of historical time.

Space, Time, Position

> On close examination, mainstream sociology turns out to be an ethno-sociology of metropolitan society. This is concealed by its language, especially the framing of its theories as universal propositions or universal tools.[29]

So, we have travelled with the ontological challenge of the sea to the critical cut of sound and the artwork: both evoke an interruption that propel us beyond the securities of territorial imperatives and deliver us into another space, heterotopic and invariably marginalized in the historical and political accounts of modernity. The perspective that arrives from the heterotopic site of the sea and the artistic interruption in representational reason provokes the freedom for a critical piracy that raids the self-assured stability of a thinking grounded in the provincial immediacies of a unique locale and language. This is to suggest an idea of history, profoundly indebted to the critical work of Walter Benjamin, in which knowledge, sustained by the search for new beginnings, proposes history not from a stable point, but via a movement in which the historian emerges not as the source but as the subject who can never fully command nor comprehend his or her language.

We are called upon to navigate languages, currents and conditions not of our own making. This is also the post-humanist confirmation that what we see does not commence from the eye, but from the external light of the world that strikes it. In the same key, it is not we who research the past, but the past that researches us.[30] This is to engage with a history composed of intervals, irruptions and interruptions. While the utopic promises an unrealizable consolation in the seemingly transparent emptiness of the present, the heterotopic promotes a disturbance, an interruption, sustained in the heterogeneity of times, rhythms and spaces, in the multiplications of modernity snapping the links in homogeneous understandings. There is no temporality or space – of modernity, of identity, of history – without other times and spaces, without the tempos and spaces of others. This is a form of knowledge production; one should say a political practice that refuses to be conscripted to a unique and unilateral understanding of modernity. It crosses the spaces of academia and institutional authorities without being inherent to them; it is a heterotopic knowledge.

Undoing the Hegelian inheritance of the presumed match between European history and reason, whereby the history of the world is synonymous with the 'reason of history',[31] is to chart the implications – critical, historical, political – of the uncoupling of modernity from Europe.[32] It is to chart modernity and its tempo-spatial coordinates on a planetary chart of entangled histories. If Europe without its 'other' is inconceivable then it is that relationship, which since 1500 has acquired sense in its global extension, and not a particular location or 'source' that provides the critical matrix. It then becomes not a question of others catching up, developing and mimicking a European measure of the world, but rather of understanding the historical and cultural centrality of that repressed and refused world to the making of Europe and the modernity it presumes to own. To transform this repressed relationship into a measure of Europe is to draw critical rigour from what exceeds its languages of knowledge and power. This, to return to the Hegelian inheritance, is to propose a history that does not simply accommodate what previously was rejected and ignored. It is rather to transpose that provincial, European inheritance into another space, signified in a lexicon of sense characterized by the planetary reach of the colonial configuration of the globe. In other words, this is not about a modernity sending out waves from the metropolis towards the periphery, seeking to harness the latter to the linear inscriptions of progress. It is rather a constellation or network, composed of shifting nuclei and relations in which distanced cultures and events can resonate in unsuspected proximities: the slave revolt in Haiti and philosophical pronouncements on the master-slave dialectic in Berlin; algebra, the mathematical meaning of nothing, the zero, in India and the Islamic world and the foundations of Italian mercantile banking. It is to consider those silent holes in the web of a planetary modernity that sustain the mesh.

If this all leads into a territory characterized by what Gurminder Bhambra in *Rethinking Modernity* calls 'connected histories',[33] its also drops us into unsuspected and uncharted waters. Learning to float, rather than seeking immediately to drop anchor or seek the shoreline, is to except exposure to a world that is not merely ours to author and authorize. Charting this space, the prevailing topologies and spatio-temporal continuum of power, formed in the exploitative flows of economic, political and cultural life can be cut to disseminate deviating discontinuities. Here the heterotopic is not only about the removal from life (the prison, the asylum and their critical negativities) but also about the deepening and extension of its possibilities (the journey, vulnerability, trans-disciplinary and trans-national knowledge). This cut or exit does not lead to an 'outside', but rather to another contemporaneity. It remains 'within' the materialization of planetary possibilities, proposing 'lines of flight': 'dub' versions of consensual reality. If culture today is increasingly disciplined by the dynamism of economical and political valuation, the seeming transparency of this self-referring bubble floats in a space where it is increasingly unable to sense what escapes and exceeds its algorithms. There continue to exist alternative accountings of the world that unplug the digital archive, crack the screen and twist the existing topologies of power into the multiplicities of the post-colonial city and the heterotopic spaces of planetary communalities that are not only capitalist in intent and extent.

This is to suggest a worldly modernity that is always under way and susceptible to unlicensed winds and currents. This is a modernity that seeds a discontinuous history; one that is out of joint with the synthesis required of an epoch that seeks only the self-confirmation of its will. The maritime archive, and the cuts induced by artworking and sonorial inscriptions, provoke and provide a refutation of that narrative. Opting to travel in an uprooted modernity that claims no unique source or history is to propose further critical beginnings.

8 'L'ASILE FLOTTANT': MODERNIST REFLECTIONS BY THE ARMÉE DU SALUT AND LE CORBUSIER ON THE REFUGE/REFUSE OF MODERNITY

Diane Morgan

A boat is a floating piece of space, a place without place, which lives by itself, which is closed in on itself and that is at the same time exposed to the infinity of the sea.

M. Foucault[1]

Nowadays one cannot conceive a utopia that does not address itself to nomads, peoples and individuals, to the homeless, to the excluded.

R. Scherer[2]

After the First World War a concrete barge made its way up and down the Seine between Rouen and Paris. It was called the 'Liège' and its mission was to supply the French capital with English coal. When it was decommissioned in 1929, the *Armée du salut* (the French Salvation Army) bought it with the aid of a donation from Madeleine Zillhardt, the recently bereaved long-term companion of the painter Louise Catherine of Breslau. Thanks to a generous donation by the Princess Singer de Polignac and a sustained campaign for public subscriptions by the 'Salutistes', the boat (renamed 'Louise Catherine') was converted by Le Corbusier into a 'floating asylum' for 150 unfortunates 'without an address, without rest, without a hovel' (see Figure 8.1). This, at first sight, rather improbable collaboration of two very different movements – Victorian Christianity and iconoclastic modernism – was the first stage of the larger 'City of Refuge' project (see Figure 8.2).[3]

Figure 8.1: The 'without slum-dwellers' [*les sans-taudis*] make their way towards 'the floating asylum, the Louise Catherine' on the Seine; *En avant*, 8 June 1926, drawing by André Labarthe.[4] Reproduced with permission of Société de l'Histoire du Protestantisme Français ©SHPF, Paris.

The standard reading of this collaboration would probably be that it was made possible by both movements' shared faith in the principles of 'social engineering'; i.e.

they both espoused the ultimately dystopian and/or reactionary values of discipline, minimalist austerity, mental and physical hygiene. It would be pointed out that the 'floating asylum' was no neutral, ideologically-free world and that, whilst offering a temporary safe haven from the ravages of capitalism, its guests were objects of charity, to be retrained and wherever possible reinserted back in that same society as Christian soldiers with the aid of a doctrinaire modernist aesthetic.[5] However, my argument will be that this project is socially, and even politically, far more challenging than one might expect. As such it – both the boat and the wider City of Refuge project – is indeed a 'heterotopia', i.e. a space of 'contestation'[6] which can unexpectedly open up a plethora of issues that are still pertinent to us today namely: the place of homeless people within our cities, our relationship towards them and our complicity with an economic system that produces their social exclusion.

Figure 8.2: The 'Cité du refuge' and its various operations: 1. The Cité itself; 2. the House of the Mother and Child; 3. the House of the Young Man; 4. the 'floating asylum'; 5. Paris by the sea; 6. the Hostel for the Liberated Prisoner in the [Guyanan] Penal Colony; *En avant*, 24 May 1930.[7] Reproduced with permission of Société de l'Histoire du Protestantisme Français ©SHPF, Paris.

The barge *Le Liège* started out as a non-space; a mere container whose purpose was to be a vast mobile, temporary storage space for raw material. It was a means to an end. This purpose served, it was saved from destruction by being converted into a well-designed living space as *La Louise-Catherine* for the socially excluded. It became an end in itself. Quantity became quality as its sheer size became habitable. Now functioning as a safe and hospitable haven to urban marginals, it was transformed into a completely different sort of space, an intrinsically vital one for those just managing to survive in the city. This radical change was itself enabled by a series of creative conversions and fortuitous encounters: the sum of money bestowed by Zillhardt, which put the barge-project into motion, had been acquired by profitably selling a drawing, which she had come across by chance, to an art dealer. When handing the windfall over to the *Armée du salut*, the grieving lesbian lover, herself no stranger to the difficulties and perils occasioned by social exclusion, explained that she was desirous of helping 'those who live under the bridges at night'. Indeed, once they had left the bright lights of the boulevards, the destitute not only descended to the darkness of the river banks, but often risked heading further, to the cold depths of the Seine itself, 'never again to return to street level' (see Figure 8.3).[8] The floating asylum was thus a beacon of hope, offering a space of hospitality and giving time to those who had felt that their time was up. Zillhardt's vision of a floating asylum was an attempt to reterritorialize the city by reinscribing what was once a zone of condemnation as a hospitable space, however temporary that might be.

The barge 'Louise Catherine' presents us with a constellation of topics and attendant issues that are 'heterotopian': water and floating vessels *per se* destabilize the habitual perceptions of the terrestrially-bound. Both remind us of the artificiality and temporariness of definite and fixed demarcations; they remove the ground from under our feet.[9] The barge was intended to act as a mirror, reflecting back to a hardened city what it did not want to see, forcing a focus on the excluded and abandoned, obliging the unfeeling to face up to their social responsibilities.[10] It was also a Noah's Ark offering a means of survival to those otherwise exposed to the hostile elements.[11] It was certainly unlike one of those unseaworthy vessels, owned by unscrupulous and exploitative human traffickers, which regularly plunge desperate asylum seekers to their death.

Water as Reality and Metaphor in *En avant*

In the *Armée du salut's* weekly newspaper, *En avant,* the pool of associations that colour our relationship to water is extensively drawn on to highlight the significance of the 'asile flottant' within the city of Paris:

> It is a beautiful thing to see a boat moored in a big port. It speaks to you of long journeys, of distant and mysterious places and adventures. But if we often pass close by it and it is still in the same place, we start to find it a bit ridiculous and absurd, like a boastful person who is always talking about Africa without having left his native village. Near to the *Ponts des Arts*, a large barge is to be found, immobile throughout the winter. This barge is neither ridiculous, nor absurd. It knows more stories than if it had frequented all the ports of the world.[12]

Figure 8.3: Desperate, the social outcast is tempted to put at end to it all by drowning himself in the river; *En avant*, 9 August 1930. Reproduced with permission of Société de l'Histoire du Protestantisme Français ©SHPF, Paris.

The 'floating asylum' together with the other refuges of the *Armée du salut*, brought together wanderers with rich stories to tell, stories that were no doubt often painful and tragic, but also fascinatingly exotic for those living comparatively 'normal', conventional, sedentary lives (see Figure 8.4).

Figure 8.4: The banquet of 800 seatings offered to those 'without a slum' [*les sans-taudis*] of Paris; *En avant,* **7 January 1928. The photograph bottom right featured in an earlier article announcing the opening of a women's refuge in the rue du Saint Sauveur, Paris 2ᵉ, with the caption 'this poor creature has not slept in a bed for the past fifteen years';** *En avant,* **11 June 1927. Reproduced with permission of Société de l'Histoire du Protestantisme Français ©SHPF, Paris.**

Whilst the floating asylum functions as a reassuring 'lighthouse' or 'haven' for those otherwise at risk of being lost in 'the immense ocean of misery and despair', it also created destabilizing effects on the city as a whole.[13] Or rather, it

Figure 8.5: The Louise-Catherine at night; *En avant*, 8 March 1930, photograph by André Kertesz.[19] Reproduced with permission of Société de l'Histoire du Protestantisme Français ©SHPF, Paris.

is precisely because it ascribed a homely place to those who do not usually have the 'right' to one, that the barge obliged the more official and established city to rethink itself. At least that was the Salutistes' intention: the 'tramp's barge' that was moored during the winter months 'in the very heart of Paris', is presented as a poke in the eye of well-dressed and comfortably housed Parisians.[14] The Salutistes take pleasure in describing the scene:

> For a few days now, those strolling along the quayside of the Louvre when it is time for Paris be lit up, look at the Seine to see it become iridescent and spangly. To their surprise, these flâneurs, for the *bouquinistes* are still open, notice downstream from *le pont des Arts,* that a renovated barge is also brighly illuminated[15]

With its attractive modernist glass roof, the boat is easily mistaken by the *mondaines* for a fashionable party venue (see Figure 8.5). They are therefore 'stupified' to see that, instead of 'ornamental ladies and men in smoking jackets' making their way to towards its beacon, there are 'poor wretches, ragged, chilled to the bone, tousled and dirty, bent-backed, who, dragging their shoddily-shoed feet towards the river, look as if they want to drown themselves'.[16] The Salutistes vividly situate their particular clientèle right in the midst of the city:

> After all the tiring efforts and sufferings of a winter's day, these pickers-up of cigarette butts, these openers of doors, these sandwich-men, these load carriers at *les Halles* market, will ask for a shower, a bed, a meal, and above all for moral support from this hospitable barge that the Salvation Army has specially fitted out for them instead of spending the night stretched out under a bridge, or huddled up in a recess or doorway.[17]

Once welcomed on board the Louise-Catherine and comfortably settled down: '[the] water that gently laps against the sides of the barge, will rock them in their sleep, murmuring to them that everything has it end, even misery'.[18]

For Christmas 1928 the *Armée du salut* served meals to eight hundred down and outs (see Figures 8.4 and 8.6). They claimed that extending festive hospitality to this assortment of people, to these *misérables,* is exactly what gives us a 'noble idea' of what society both is and should become: '*c'est ça qui donne une fière idée de la société*'.[20] The heterotopian City of Refuge is to become a project for a society as a whole. As such, the heterotopia it creates are also utopian: rather than just 'suspect[ing], neutrali[sing] and invert[ing]' other actually existing places, their intention is to 'efface, neutralise and purify' current societal emplacements.[21] Boats invariably invite one on a long journey out of oneself, towards unknown 'exterior space[s]'.[22] What needs to be flushed out of ('purified' from) society is injustice, and this is to be achieved by injecting into the main body of society, those – barely known, hardly encountered – who have been up to now social outcasts. *L'Armée du salut* considers that the most destitute, including, or *especially,* criminals, have something positive to offer society.[23] They therefore in effect, advocated and even practiced, what Nietzsche called, societal 'ennoblement through degeneration'.[24]

Figure 8.6: Those who were invited to the Christmas meal; *En avant*, 7 January 1928. Reproduced with permission of Société de l'Histoire du Protestantisme Français ©SHPF, Paris.

En avant!: *L'Armée du Salut*, More than Evangelicizing Tambourine-Bashing.

'The Salvation Army', these three words used to unleash gibes. A red jersey or a Miss Heylett hat with Salutiste coloured ribbon sufficed for laughter to explode and cooked apples to fly. The times have changed. The Salvation Army now acts as a password and introduction into the most diverse milieu, ranging from the narrow attic room to the sumptuous hotel of rich bankers, of powerful industrialists or of the old families of the wealthy *faubourgs*.[25]

The Salvation army revives the propaganda of early Christianity, appeals to the poor as the elect, fights capitalism in a religious way, and thus fosters an element of early Christian class antagonism, which one day may become troublesome to the well-to-do people who now find the ready money for it.[26]

A focus on the 'heterotopian' spaces of the Salvation Army, at least on the form it took in France during its years under Albin Peyron's command (1917–35), can challenge stereoptypical ideas about this organization. The movement could be regarded, not only as less traditional than one might have supposed, not only as reacting to the negative social conditions brought about by urbanization, but also as militating for controversial causes and embracing the potentially utopian aspects of modernity and modernism. Indeed, as the very title of '*En avant!*' suggests, the organization aimed to lead the way for social change in several ways (see Figure 8.7).

Paris was the first port of call on continental Europe for the three young English female missionaries dispatched from the Head Quarters in London in February 1881 to set up a sister organization.[27] However, by 1917 Albin Peyron was the head of the French office and under his – together with Blanche, his wife's – leadership, it developed its own distinctive style and specific social missions.

One of the earliest projects of the *Armée du salut* was the People's Palace (*Palais du peuple*), for young men opened in 1912, and extended with an annexe (plans drawn by Le Corbusier), in 1926. Together with the Women's Palace (*Palais de la Femme*) opened in 1926, this foyer was incorporated into the larger City of Refuge project. The larger complex was conceived of as a dynamic and efficient 'sorting station or to be more precise, a turntable' of the socially disadvantaged and excluded. This very modern 'organ of distribution' was to be continuously communicating with the various social services of the *Armée du salut* and public services, aided by the use of the telephone.[28] These concerted and coordinated forces are what produce effective action.

Figure 8.7: 'A city where the poor will find his way', *En avant*, 16 November 1929. The caption reads 'God has prepared a City for them'. Note the art déco heading. Reproduced with permission of Société de l'Histoire du Protestantisme Français ©SHPF, Paris.

Figure 8.8: 'During the week of renunciation, please no abstention. The Factory of Good must continue its work', *En avant*, 24 October 1931. Reproduced with permission of Société de l'Histoire du Protestantisme Français ©SHPF, Paris.

The City of Refuge presents itself as the multi-sited place where 'the wheels of the human machine' that have been 'worn out by life' will be serviced (see Figure 8.8).[29]

However, the *L'Armée du salut's* operations were not just about fixing the human spare parts of the societal machine with the result that those 'saved' were fed back as fodder fit for consumption by labour market. Indeed, this type of 'charitable' injection, that can not swing society into a different transformative mode of operation, was precisely the approach that they criticized:

> This is no generous and utopian dream to be realised any old how in little doses, following the charitable fancies of a few devoted hearts. The precise duty of town councils and the State is to do exactly what the Salvation army does, but more extensively and on a bigger scale.[30]

The *Armée du salut* aimed much higher than solving a social problem. Its grander vision could strike those fatalistically resigned to a supposedly 'natural' state of affairs as 'mad'. However, despite his apparent dismissal of 'utopian dream[s]' (expressed above), Peyron fully recognized the powerful potential of utopian 'craziness'. He proclaimed that:

> the nomad, the vagabond, he who is tired, he who is desperate, he who is dying of hunger, he who is without a slum, he who is without a faith, he who is place-less ... all of them will be able to come at any hour with the certitude of being welcomed, fed, given a bed, clothed, comforted, advised, cared for. This *mad dream* will cost 10 million francs and it begins to be translated into reality.[31]

Once realized, the City of Refuge was in turn to become a model for society as a whole to emulate with a view to changing its internal composition and organization. Using language worthy of Charles Fourier, they describe how:

> [this] city will get built, the movement imposes itself as it offers a final synthesis of all the tendencies of goodness, of all the good works, that were to this day not concentrated and that will now find in this liberating formulation their rallying point.[32]

En avant does employ the language of recuperation and recycling to depict how the Salutistes convert individuals into valuable substances (e.g. 29/03/1930 and again see image eight). However, just as much onus is placed on social critique and societal transformation. The concentrated energy and 'irresistibl[y]' dynamic 'movement' of the evolving multi-sited project takes society with it, radiantly inspiring 'ever growing interest' in its projection of a better future (see image nine).[33] The public's enthusiasm for the project is augmented by the force of the Salutistes' own exampleof 'disinterested' commitment to a good cause.[34] This exemplary behaviour is then emulated by the public, who in turn are increasingly enthused by their own generous mobilization.[35]

Figure 8.9: The radiating City of Refuge complex; *En avant*, 6 July 1929. Reproduced with permission of Société de l'Histoire du Protestantisme Français ©SHPF, Paris.

The project becomes more and more realizable as people, each according to his or her means, send in contributions:

> Everyday numerous letters arrive containing donations. These contributions are frequently accompanied by a few touching words. Vast waves spend themselves against dykes made up of grains of sand; similarly, the numerous donations, whether they be small or big, will permit the construction of the City, a dyke which will cause the waves of misery to be transformed into elements of life.[36]

Drawing its metaphorical energy from the hostile nature of water, *En avant* describes the cumulative resistance, akin to a dyke, to the current state of affairs, whereby the destitute are just abandoned to their misery. However, the public is not just expected to delve into their pockets to alleviate social ills. Venturing beyond a reactive stance, they are also expected to open their minds to a conception of a future society that hospitably includes the social marginalized as different types of people whose condition is to be improved, most certainly, but who will be also allowed to be as different as they want to be. *En avant* gives the public a foretaste of this community to come, a society whose spaces will henceforth include this eclectic multitude, that will incorporate those who were previously conceived of as being just one (undesirable) homogenous mass. Describing the clientèle who frequent the People's Palace, they write:

> [There are firstly] the workers, a fair number foreigners, on causal contracts, or paid parmoniously by some thrifty business. These come back home, harrassed, beaten down. They eat a bowl of soup and vegetables and then go to sleep. All for 5fr 25. They pay from one day to the next, or for the week; they are regular customers but as soon as their situation improves, they go a hotel closer to their workplace.
> The second category consists of employees, representatives, professors and *déclassés*. There are some who stay for a night and then disappear, and those who have been here [rue des Cordeliers] for some time and consider themselves to be part of the fabric of the place.
> Those who were once rich, now poor; ruined gamblers; the unhealthy; the sickly; retired people on small pensions; inconsolable widowers; serfs; eccentrics; misers; as well as prodigies who stay a week or two as they cannot maintain their lifestyle until the end of the month. Foreigners, well, foreigners without papers; professors in their language; globe-trotters on foot, on bicycles, on scooters, on their hands, or going backwards; fakirs; students; clergyman; slavic popes; journalists from all the Near Easts and engineers from all the centres of Europe; sons of families and those without families; communists; fascists; anarchists etc. but all with papers, felt hats, beautiful conversation, polished manners and a genuine smile.
> And in the evening gathering in the large reading room, all these people talk, smoke, read, dream or doze in the heavy atmosphere created by two hundred breaths infused with the incense of tobacco, without angry outbursts, without a discussion or a argument that transgressses the limits of puerile and honest civility.[37]

These passages give us a description of the heterotopian community that existed at the People's Palace and indeed probably still exists there and in other refuges today. They also give us a vision, one we could call 'utopian', of what society as a whole might become. The community is characterized by a manifold of personalities, as such it is far from grounded in any homogenizingly identitarian doctrine. It is also at least partially modelled on self-government as we are told that in this hospitable house 'a sort of democracy' is at play: 'those who arrive wretched [*misérables*] come under the influence of those who have picked themselves up'.[38] Also to be noted is that the very language of this colourful taxonomy of the community's components is 'heterotopian' inasmuch as it: [breaks up] all the ordered surfaces and all the planes with which we are accustomed to tame the wild profusion of existing things and threaten[s] with collapse our age-old distinction between the Same and other'.[39] The categorization between the groups consists of shaded differentiation (e.g. 'globe-trotters on foot, on bicycles, on scooters, on their hands, or going backwards', 'the unhealthy, the sickly') and juxtaposed extremes ('communists, fascists, anarchists'); the overall picture is a medly of different, yet associated, personalities and groupings. This community poses us a challenge: it is a heterotopian community that we are being asked to accommodate in the future 'order of things'.[40]

Figure 8.10: 'Our pupil officers contribute to the victory. Step after step the troops attack the mountain peak', *En avant*, 11 April 1931. Reproduced with permission of Société de l'Histoire du Protestantisme Français ©SHPF, Paris.

The ongoing march towards the realization of more of the City of Refuge project will take the Salutistes not only higher and higher towards their goal, but alsofurther afield, taking the public with them (see Figure 8.10). Soon the force of its example will propagate itself beyond the capital outwards across the whole nation: 'the whole country will be renewed, renovated, put into action'.[41] Ever true to the nature of the water that irresistibly draws us outwards, towards the wide oceans, the *Armée du salut* then push the boat out yet further, beyond the French mainland: 'The City will radiate in the provinces and even further, across the ocean, towards those far shores where the wretched [*les malheureux*] wait for a little pity, a little justice, a little love'.[42]

The City of Refuge, that was initiated with the 'floating asylum', stretched right out to the penal colonies [*les bagnes*] of French Guyana.[43] This outreach project was a distinctive sign of the *Armée du salut's* committment to a controversial, or at least not evidently popularist cause. In one of his weekly 'Chronicles' entitled 'They are waiting for a homeland [*une patrie*]', Peyron himself stated: 'The City of Refuge and the penal colonies will never leave my thoughts'.[44]

The French Guyanan penal colonies [*les bagnes*] were originally invented in 1795 as a dumping-ground for political prisoners during the French revolution. Thereafter, as from 1854, they were used for criminals. Charles Péan undertook a mission for the *L'armée du salut* to investigate these hellish places in 1928. He published his findings in *Terre de bagne* (1933) and the *Armée du salut* used the book to bolster its already intensive campaign wherein the fate of such prisoners was accorded the same importance within the City of Refuge project as as the plight of the destitute in France itself.[45] What was especially shocking about these penal colonies was that 'liberation' turned out to be even worse than imprisonment.[46] When incarcerated and subjected to forced (and 'Sisyphean') labour, one at least was given some degree of food and lodging.[47] Once the prison sentence had been served, prisoners were just turfed out, with no means of survival at all. In effect they had been condemned to a slow death. By not protesting against this practice, the public was in effect committing a 'social crime'. Hamp bluntly makes his point: 'When it comes to an individual crime there is a guilty person. When it is a question of a social crime no-one is innocent'.[48] Not only that, Hamp accuses the public of actively parasiting and profitting from this criminality. True to their practice of rewriting the spaces 'we' often take for granted (as discussed earlier), the *Armée du Salut* again 'injure the eyes of the French people' by forcing them to see just how their beautiful sites of 'culture' are constructed on the back of the 'barbarism' that is the penal colonies:[49] 'Nice would never have become such a popular city of entertainment had the penal colony had not been transported to Guyana'.[50]

The *Armée du salut* unequivocally condemned the fully legalized practice of the penal colonies, as well as challenging the very idea of The Criminal. They pointedly interrogated their readers' assumptions about this 'type' of person: 'What is a criminal? Is it a man within whom everything is criminal? Or is it a man who at one moment or another committed crimes?'.⁵¹ The message is clear: it is society's duty to save what remains of the human in these (often dangerous) people, despite – or rather *precisely because* – they are 'those who have fallen lowest', because their 'faces have been eaten away by the cankers of their hearts', because they are 'disgusting '; like 'purulent sores'.⁵² Acts of charity are to be carried out in the face of such ugliness and of such wickedness, without the expectation of recognition and gratitude.⁵³ It is precisely the exposing of onese lfselflessly('disinterestedly') to such a 'heterotopian' situation, by embracing the rubbish (*déchets*) of society, that society can be radically transformed.⁵⁴ Instead of entrenching those who had themselves violated the law further into their asociality by in turn committing an injustice against them, the *Armée du Salut* advocated that society should reconceive itself as an 'ally' of such miscreants, just as it had, closer to home, espoused the cause of those who are 'a bit frightened, dirty, sometimes pitiful [*lamentables*], others [who are] magnificent in their beggarliness [*magnifiques dans leur gueuserie*]'.⁵⁵ By so doing, society could find itself capable of performing 'miracles'; it could find itself facilitating that which it previously would have dismissed as 'utopian'. Indeed, 'great transformations' can take place in these people if they are given refuge.⁵⁶ These criminals, some of them murderers, can in turn become life-savers, 'snatching back from the abyss and from death, other shipwrecked persons who were about to be swallowed up by the storms of life'.⁵⁷ Cities of refuges, as described in the Bible, protect murderers against revengers, providing them with spaces for atonement.⁵⁸ As part of a project for society as a whole, they are seen as working towards the materialization here on earth of the future City of God.⁵⁹

Le Corbusier and the Radiant Future

> the earth is not a prize to be won in a race ... there is a place for everyone under the sun.⁶⁰

> The end of the old civilisation has come; the face of the earth will be renewed under a new sun.⁶¹

> The whole country will be renewed, renovated, put into action: a place not only decent but radiant for everyone under the sun.⁶²

> The whole country will be renewed, renovated, put into action. It will reach that 'why' for which one makes revolutions. A place that is not just decent but radiant for everyone under the sun.⁶³

As has been already stated, Le Corbusier's 'floating asylum' is rarely discussed even if boats feature widely not only in his architectural projects, but also in his writings.⁶⁴ Indeed, he esteemed that landlubbers ['*les terriens*'], who are far too set in their traditional ways of living, had a lot to learn from the 'pure, precise, clear, clean, healthy' architecture of ocean liners.⁶⁵ The neglect of his connection with *l'Armée du Salut* is even more remarkable given his continued collaboration with them, after the Louise-Catherine barge conversion, on the bigger *Cité du refuge* project. He also explicitly refers to them in *Sur les quatre routes* (cited earlier). Le Corbusier goes so far as to present *L'armée du salut* as *the* organization that can teach the public how to live in the future. Such a recommendation is no negligible affair coming from someone who so often presented himself as representing the future. According to Le Corbusier:

> The Salvation Army knows how to detect the poverty-striken (*les misères*) and identify decent people (*les braves gens*) ... Then, being an attentive sister, it will teach them how to live in their houses, as *knowing how to live* is a technique, and how to live in the lodgings of the new spirit ... requires education.⁶⁶

The *Armée du Salut* is here presented as an organization that can keep pace with the spirit of modernism, and help people adapt to the new conditions. In the earlier book, *The Radiant City* (1933), Le Corbusier also refers to the *Armée du Salut*'s community work, praising its ability to make a 'direct public appeal' in contrast to other, more official and therefore ostensibly more powerful, institutions.⁶⁷ The impression is given of someone who is keenly aware of society's injustices. Highly critical of those in power, he explicitly takes the side of the *Armée du Salut* as the effective representative of ordinary people against the dominant order. However, it is not only sympathy for the plight of those less fortunate that motivates his analysis; the combined forces of the *Armée du Salut* and ordinary people, including the destitute, are also presented as emboding the vital values of the (utopian) future.

However, elsewhere in *The Radiant City*, we encounter a different side to Le Corbusier, one that can easily strike us as far less 'socialist', as far less concerned with the details of ordinary lives, as far less sympathetic, indeed as downright sinister and, one could say, 'dystopian'.⁶⁸ In the above quotation, his analysis of the disease afflicting contemporary society of 'moral corruption, embezzlement, betrayal of trust' complemented well the *Armée du Salut*'s active investment in heterotopian spaces for the poor in the name of a utopian future for all. However, Le Corbusier's critique of materialist society then associates itself with anti-Semitism, the extreme right and Pétainism.⁶⁹ His declaration that 'we must pull things down ... And throw the corpses onto the garbage heap' strikes a dissonant chord with the *Armée du Salut*'s investment in 'refuse'.⁷⁰ He appears out of tune with the social commitment to, and creative revalorization of 'beggarliness' ('*gueuserie*') as documented in *En avant*.⁷¹ However, such discordance should not really be of surprise to us given the speech Le Corbusier gave at a meeting of 'Le Faisceau', the first French fascist group, on 20 May 1927 when they were inaugurating their new headquarters in Paris 10ᵉ.⁷² In his article 'The New Stage of Fascism. The Way to Success through Poverty: Three Symbolic Days Charged with Meaning and Hope' for their newspaper *The New Century*, Georges Valois, explains how Le Corbusier's speech about the modern city expressed 'the profound thoughts of fascism, of the fascist revolution'.⁷³ Just like the 'genial' Le Corbusier who paid them a 'great honour' by attending their meeting, fascism also values urbanism; it recognizes the need to conceive the city as whole and to 'coordinate forces' within an overall plan. This urban planning breaks with the haphazard constructions of the past. It is to be carried out in the name of the poor who have for far too long been subjected to the consequences of shoddy housing. Valois instrumentalizes the slum-dwellers, using them to support the fascist cause, when he proclaims that:

> [w]e can no *longer* accept that anyone should live in extreme poverty [*la misère*]. This engagement presupposes a new political, economical and social organisation.⁷⁴

Such associations rock the boat; it is as if our floating asylum for the poor is being hijacked by the wrong sort of people...

Conclusion: Heterotopia, Utopia and Dystopia

In this chapter I have analyzed the 'asile flottant' not only as a heterotopian space within the city of Paris, but also as a component of the extensive and dynamic City of Refuge project. This second aspect of the barge's function meant that it was contributing to a wider vision for social transformation, one that could be called global utopianism, intent on changing the shape and quality of all society 'under the sun'. However, I also considered how this project was haunted by ugly dystopian features, in the form of the far right.⁷⁵ To conclude, I wish to consider how Foucault's idea of heterotopia strangely replays many of these issues, albeit in a lighter mode.

Earlier I made the point that, in the 'Of Other Spaces' essay, Foucault distances 'heterotopia' from utopian thinking, whilst also enigmatically suggesting that, at least in the form of a mirror, they could have a 'mixed joint experience'.⁷⁶ He reduces the role heterotopias play to just 'suspect, neutralize and invert' other actually existing places, instead of having the potential to 'efface, neutralise and purify' current societal emplacements.⁷⁷ He restricts utopias to 'unreal spaces'. The repercussions for our analysis would be that the *Armée du Salut*'s initiatives would be simplistically reduced to a mere reaction to the negative social conditions brought about by urbanization, instead of appreciating their various refuges as meaningful attempts to bring about effective change for the future. Indeed, one of the many 'tribulations' of Foucault's idea of 'heterotopia' was its identification with a negation of history and thereby of the very possibility of radical change ever taking place.⁷⁸

The proposition that the utopian potentiality of 'heterotopia' is scaled down by Foucault between the two versions of the text resonates with another shady story: I am referring to the occluded 'debate' between Foucault, Marcel Gauchet and Gladys Swain on the subject of madness. Looking back after Swain's death at their research on the history of psychiatry, work that in effect dismissed Foucault's theory of 'the great confinement' of the 'mad' as a 'myth', Gauchet questions why the philosopher never responded to their serious allegations against him of oversimplification and academic lack of rigour. Gauchet and Swain had painstakingly documented how inaccurate it was to present asylums just as repressive heterotopias, even though their dystopian features were undeniable. They insisted on how asylums were also part of a movement that paradoxically could be considered committed to 'utopia as praxis'. Gauchet explains:

> [Asylums] embodied a moment of belief in the possibility of creating *ex nihlio* in a vacant space outside of society, another social microcosm, that is at once wholly specific (i.e. that only responds to its own rules), and yet fully social.[79]

Gauchet and Swain make it clear that the segregation of the 'mad' in asylums went hand in hand with a belief that those who had previously been locked into individual isolation within the 'community', were capable of socialization. Asylums therefore constituted a recognition that these otherwise excluded people could be integrated into a collective body. Foucault's silence on this well-documented issue is taken as a sign of his will to institutional power. His reputation and authority is constructed on the back of the 'myth' of the silencing of other voices in those other places.[80]

However, the intrigue surrounding the question of heterotopia and its relation to utopianism continues further. Gauchet has since announced his return to 'normal politics' and criticized Miguel Abensour for his unrealistic utopianism. Gauchet's dismissive categorization of Abensour as a left-wing 'revoltist' has been read by the latter as a confirmation of a 'veering towards the right'. In his open Letter of a 'revoltist' to Marcel Gauchet, who has converted to 'normal politics', Abensour counter-attacked as follows:

> 'Normal politics' is the other name of the hatred for alterity, of all social others. In a word, it is the contemporary face of the hatred of utopia. Once one is resigned to 'normal politics', a world establishes itself where all utopian divergence [*l'écart utopique*] has disappeared forever.[81]

For Abensour, the anti-utopian stance of Gauchet prepares the way for a 'dystopian' future: 'This closing of society in on itself, this 'soft' form of [La Boétie's] voluntary servitude, besides ignoring the persistance of utopia throughout history, condemns human society to the repetition of the same.'[82]

Such a closed society, with its assumption of homogeneity, normality and conformity leads directly to totalitarism. It can therefore be seen that documenting the 'persistance of utopia throughout history', for instance in the form of the complex heterotopian space that was the 'floating asylum' with its heterogeneous, nonconformist and transient population, can have far-reaching political consequences for the future.

Acknowledgement

My warm thanks to Roger Palmer for embarking me on this voyage of discovery with '*l'asile flottant*' and to the librarians at the *Bibliothèque de l'Histoire du Protestantisme Français* in Paris for making *En avant* available to me.

9 ZOOHETEROTOPIAS

John Miller

The life and work of the Belgian zoologist Bernard Heuvelmans were notably unorthodox. Completing a doctorate on the teeth of the aardvark at the Free University of Brussels in 1939, he escaped four times from Nazi prisoner of war camps to make an unlikely living as a comedian and jazz singer in post-war Paris. It is his contribution to an esoteric branch of science, however, or perhaps more accurately pseudo-science, that constitutes his most prominent legacy. Heuvelmans' zoological interests, as they were to develop throughout a prolific writing career, were concentrated on forms of life at the margins of conventional natural history, beings, he explained, 'the existence of which is based only on testimonial or circumstantial evidence or on material proof judged insufficient by some'.[1] His research was painstaking, but dealing with animals more usually confined to the territory of the mythic and the fantastic inevitably placed him at a distance from institutional science. Perhaps unsurprisingly, he would never gain a formal university appointment. Nonetheless, he remained committed to the study of ambiguous animals (including, most famously, the yeti and the abominable snowman) and to the possible rediscovery of creatures 'universally thought to be extinct'[2] – dinosaurs, the mammoth and the New Zealand moa among others. By the time of his death in 2001 aged 84, Heuvelmans was widely honoured as the father of cryptozoology, a term he coined in the 1950s (although he later acknowledged that it had already been used by another creature seeker, the Scottish adventurer Ivan T. Sanderson).

If Heuvelmans' work received a measure of recognition, it was nowhere near the recognition he felt it deserved. Cryptozoology continued to be routinely scorned by the scientific establishment as the province of cranks, hobbyists and fraudsters; conversely, the scientific establishment was routinely condemned by cryptozoologists as narrow-mindedly attached to what Sanderson identified during an ill-tempered spat about Bigfoot in 1967 as a 'whole gamut of orthodoxies'[3] that prevented the acceptance of Sasquatch *et al.*, even in the face of (to some observers) convincing evidence. Heuvelmans, accordingly, opens his 1955 book *On the Track of Unknown Animals* in combative mood:

> Most zoologists are sceptical about the possibilities of discovering new species of large animals, but some of them do not, with legitimate scientific scepticism, keep an open mind until the species is proved to exist, but categorically deny that it can possibly do so until they have been forcibly proved wrong.[4]

For Heuvelmans, the taxonomic door is always already conspiratorially closed on a wide variety of animals that are extensively attested in anecdotal evidence. Such robust rhetoric reprises a tension, apparent since the nineteenth century, between professional men of science and unpaid enthusiasts (or, as Brian Regal puts it in the most significant academic study of cryptozoology to date, the antipathy between 'eggheads' and 'crackpots').[5] Heuvelmans' lament helps to construct a recurrent sense of the cryptozoologist as an enlightened but long-suffering maverick battling against a hidebound establishment. As Regal argues, Heuvelmans 'established the trope of the free thinking, unfettered amateur monster hunter fighting against the dark forces of scientific mainstream conservatism'.[6] Indignation is the characteristic tone of much cryptozoology; but cryptozoologists often also, it seems, relished their status as outsiders.

Notwithstanding Heuvelmans' self-confidence, there are, undoubtedly, methodological problems in his work. Cryptozoology's academic marginality is less to do with its central idea that many species are undescribed by science (which remains largely uncontroversial, although there are greater problems with the very large creatures cryptozoologists often prefer) and more to do with its specific focus on or attraction to less reliable forms of knowledge. Cryptids, as these lost creatures are known by aficionados (and also sometimes, evocatively from the French, *hauntiques*), are beasts of rumour, hearsay and legend. Heuvelmans trawled through travellers' tales and journal articles looking for stories, hints even, of unexplained animal presences. A significant part of Heuvelmans' scholarship is therefore textual; his work bestrides the zoological and the folkloric, forcing the two cultures into an uncomfortable intimacy, reimagining narrative as evidence.

Such a fascination with the traces of extra-zoological animals does at least have the benefit of a long history. What is *Beowulf*'s Grendel but a cryptid *avant la lettre*?[7] Mysterious animals have been an enduring part of literary tradition and Heuvelmans' work is entangled with literary sources and influences from its inception. It was apparently early readings of Jules Verne's *Twenty Thousand Leagues under the Sea* (1870) and Arthur Conan Doyle's *The Lost World* (1912) that encouraged his life-long enthusiasm for the secrets of natural history. Indeed, *On the Track of Unknown Animals* contains several pointed references to Conan Doyle and to *The Lost World*, in particular, which evidently remained something of an inspiration. 'I do not suggest', Heuvelmans concedes:

> that we can hope to *emulate* Conan Doyle's *The Lost World*, in which Professor Challenger found on a [plateau] *all* the enormous reptiles from all parts of the world in

the Mesozoic era and even some ape men. But it is absurd to insist that there can be *no* unknown animals in such little-known country.[8]

In a further consolidation of his literary credentials, as a friend of Georges Prosper Remi, better known as Hergé, Heuvelmans served as an advisor for many of the Tintin stories, perhaps most significantly feeding his expertise in Yeti myths into the 1960 *Tintin in Tibet*. Without the security of an academic day job, Heuvelmans relied on marketing his own books to a wide audience, clearly with a good deal of success. A *Daily Mail* review noted approvingly that 'Dr Heuvelmans' original research beats any alleged thriller for enthralling excitement'.[9] His voluminous cryptozoological output, therefore, necessarily nourishes rather than shies away from the imaginative potency of mythical and semi-mythical beasts. The beleaguered Heuvelmans' determination to be taken seriously as a naturalist was always uncomfortably entwined with his interest in and contribution to a literature of the fantastic.

Importantly, Heuvelmans's attempted formalization of cryptozoology as an academic discipline in the 1950s emerges in the wake of a period of remarkable super-abundance for mythical creatures in the last decades of the nineteenth century and the first of the twentieth, an era which provides many of the keystone species of the extra-zoological imagination: not just Conan Doyle's dinosaurs in *The Lost World*, but Edgar Rice Burroughs' prehistoric fantasy *The Land that Time Forgot* (1918); also, of course, the most famous cryptid of them all, *King Kong* (1933). These outsized beasts have evoked an ever-expanding repertoire of symbolic or allegorical interpretations, as if their hyperbolic bodies require the hyperbolic literary critical practice that Roger Luckhurst identifies as 'monstrous over-reading'.[10] Psychoanalysis, in particular, dogs the cryptid's footsteps; monsters, as Regal flamboyantly asserts at the opening of his book, 'have stalked the dark parts of the human psyche, as well as forests, for millennia'.[11] But as much as cryptozoology and its explicitly fictional avatars suggest the purely fantastic, they may also be read in materialist terms as a response to environmental history; as part of a discourse of animals, and more specifically, animal habitats that emerges in the context of industrial and imperial modernity.

The lost worlds of the *hauntiques* function, I will argue, as zooheterotopias, territories which comprise, as Foucault explains his conception of heterotopia, 'a simultaneously mythic and real contestation of the space in which we live'.[12] These uncanny spaces exist (if that is the right word) outside familiar, regulated geographies and generate a particular affective experience that reflects back onto and adjusts understandings of all other sites. Specifically, zooheterotopias reveal the complex involvements of global capital with the ecological imagination and function in terms of a paradox. On the one hand, these habitats articulate the desire for spaces ungraspable by capital, for animal life beyond the reach of imperial modernity; on the other, it is the very 'outside' of capital that drives the urge to representation and appropriates this 'beyond' into the ideological regime it appears to contest. Cryptozoology, therefore, emerges as a point of tension between the possibilities of a radical environmentalism and the disappearance of conservation into economic imperatives. The central purpose of this analysis is to utilize Foucault's thought as a contribution to a self-critical environmentalism that remains vigilant of the historical intimacies between ecological discourses and discourses of global capital.

Blank Spaces/Lost Worlds

It has been consistently noted that Foucault's oeuvre contains little or nothing in the way of an explicit environmentalist or pro-animal focus. As Chloë Taylor contends, the 'few times that Foucault discussed animality or human-animal relations, animals and animality remained metaphors for humans and human experiences'.[13] Nonetheless, his work has proved influential in the formation of critical fields committed to unravelling the logics and strategies of human domination in a period of ecological crisis, most notably, as Sara Rinfret argues, through 'three key concepts – disciplinary power, governmentality and biopolitics'.[14] The conception of the heterotopia is also one with clear, if implicit, ecological resonance. Foucault's contention in the opening paragraph of 'On Other Spaces' that the 'present epoch will perhaps be above all the epoch of space' has the ring of self-evidence for environmentalists. There can be no thinking about ecology, about life forms, their interactions and the forces that shape these, without thinking about the spaces in which these interrelations unfold and necessarily, therefore, about the processes (both material and discursive) that produce, modify and reproduce these spaces. Given the way in the twentieth and twenty-first centuries (continuing processes initiated in the eighteenth century if not before) that ecological conditions are determined by economic considerations (especially in the current age of 'biodiversity offsetting' and 'ecosystem services'), Foucault's insight might now be refined: our present epoch is above all the epoch of space in relation to capital.

The movement to safeguard key areas of biodiversity is by now a well-established, if intermittently successful, aspect of green campaigning and appears (in part) as an ongoing battle against the priorities of corporate interests in a time of resource scarcity. Historically, the development of environmentalism as a spatial discourse reaches a particular intensity at the start of the twentieth century with the enthusiastic designation of national parks in response to the unrestrained environmental excesses of the Victorian period (especially in terms of overhunting and the development of large-scale colonial agriculture). Conservation is based to a large extent in the administration of natural habitats in order to prevent and redress unrestricted development and as such often depends upon an ecological bureaucratization of territory that increasingly involves a biopolitical assessment of animals as populations to be managed. As much as conservation involves a contestation of the commercial exploitation of habitats, it has also emerged as itself an economic discourse, most recently through the formation of 'natural capital' as an incipient mechanism for attaching value to nature, and before that through the arrival of 'sustainable development' as a means of incorporating an environmental perspective into the goals of global capitalism. It is clearly an oversimplification, therefore, to think of environmentalism (a phenomenon that itself includes multiple ideological persuasions) in direct opposition to capitalism. Ultimately, there can hardly be a more vital, or intractable, political question at the start of the twenty-first century than the vexed relationship between the economic and the ecological.

For all its scientific marginality, cryptozoology offers a useful lens on the way that economic and ecological discourses come to be entwined in the first half of the twentieth century. Cryptozoology too is inextricably connected to questions of space, and also (though less obviously) to questions of capital. It exists, in Marlowe's famous phrase from Conrad's *Heart of Darkness*, in a 'blank space of delightful mystery – a white patch for a boy to dream gloriously over';[15] as much a discourse of geography as it is a discourse of species. The search for lost or undiscovered animals hypothesizes not just a state of zoological possibility, but also, connectedly, insists on the endurance of fantastic habitats beyond the reach of natural historical orthodoxy. Given cryptozoology's specific focus on megafauna, the question of habitat is notably accentuated: if dinosaurs still roam the earth, where have they been hiding? Accordingly, Heuvelmans' chapter titles in *On the Track of Unknown Animals* routinely announce both creatures and territories: 'Nittaewo, the Lost People of Ceylon', 'Orang Pendek, the Ape Man of Sumatra'. At the moment when it seems the world's secrets have been exhausted, cryptozoology posits some secluded cave, remote forest or neglected island where a few last specimens have miraculously evaded the reach of science. In a section on the New Zealand moa, Heuvelmans quotes the nineteenth-century German geologist Ferdinand von Hochstetter's suggestive summary of the quintessential territories of the last survivors: 'It is by no means impossible that in *lonely and inaccessible places* live a few scarce diehards of the giant family'.[16] Cryptozoology flies in the face of a widespread supposition that the world's last secrets are gone forever. 'There are lost worlds everywhere', Heuvelmans avers in the title of the opening chapter of *On the Track of Unknown Animals*. In a period when the world appears to have become thoroughly known (in part through the development of conservation and the ever extending reach of natural history expeditions), cryptozoology insists on an enduring epistemological lack. 'The world', for Heuvelmans, 'is by no means thoroughly explained'.[17] Cryptozoology, then, may be understood both as a response to the totalizing effects of modernity and as an intriguingly resistant discourse.

There is more than coincidence to the increasing fascination with cryptids during the years that saw the formalization of national parks in the first decades of the twentieth century. Wilderness is no longer a glorious blank space; the production of habitats as objects of knowledge contributes to the emergence of a tone of mourning that is a consistent note in cryptozoological fictions. As the newspaper editor McArdle laments before the adventurers set sail for South America in *The Lost World*, 'The big blank spaces in the map are all being filled in and there's no room for romance anywhere'.[18] These lines both echo Conrad's and anticipate the conclusion of another Victorian and Edwardian novelist whose work abounds with cryptids, Henry Rider Haggard. Haggard is most notable in the annals of cryptozoology for his tales of lost gorilla gods in two later novels *The Holy Flower* (1917) and *Heu-Heu* (1924), which influenced Merian C. Cooper's and Edgar Wallace's development of *King Kong*. Writing in his autobiography, *The Days of My Life*, Haggard poignantly announces that 'with all the world explored and exhausted, I feel sorry for the romance writers of the future, for I know not whither they will turn'.[19] (London: Longmans, 1926), vol. 2, p. 91. For both Haggard and Conan Doyle, the start of the twentieth century is marked by a global aesthetic impoverishment produced in part by habitat loss (deforestation, urbanization and the rise of big agriculture – the usual suspects of environmental degradation) and in part also by the development of conservation as a practice of land management. The modern world is shorn of romance, there is no longer a 'secret place', as Haggard puts it elsewhere, 'unknown to the pestilent accuracy of the geographer' in which writers can 'lay their plots'.[20] Such a longing for romantic literary territory stimulates a consistent note of anti-materialism in the work of Haggard and others as adventure comes to require a wilderness untouched by commerce.[21]

Cryptozoology, then, is part of a yearning for romance and as such is intimately involved with loss, a connection that surrounds it with a notable irony. For all the commitment of enthusiasts like Heuvelmans, cryptozoology retains a specific investment in its own failure. As soon as a cryptid is found, it stops being a cryptid; the fascination of its secrecy is ruined by discovery. If the aim of cryptozoology is to bring secret creatures to light, there remains paradoxically an aesthetic frisson in their concealment. For all Heuvelmans' desire to taxonomize the yeti (*Dinanthropoides nivalis*, though it has never caught on), cryptozoology is structured around what might be thought of as an anti-taxonomy. It is the study of the unknown, even the unknowable that drives the quest for the *hauntique*. The term cryptozoology itself, etymologically, refers to the study of *hidden* animals. The cryptid is condemned to exist perpetually in the moment before revelation. Its natural habitat, so to speak, is the blurry photograph or shaky amateur film, premised on images which often could be anything: undiscovered species, a disguised human form or any number of discarded household objects. As an enterprise at the edge of orthodox academic practice, cryptozoology itself, like the animals it studies, evades plain visibility. Researching cryptozoology today is a tale of abandoned websites, broken links and defunct journals. The International Society of Cryptozoology was founded in 1982, but became extinct in 1996, as did its journal *Cryptozoology*.[22] It seems as if this nonconformist undertaking is unwittingly performing the kind of breakdown in or dereliction of the structures of modernity that might allow unexpected animals to retain a tenuous foothold. Both the forms and the content of cryptozoology are useful, then, in unpacking its seductive mythology of lost and hidden worlds.

To appreciate more fully the ways in which cryptozoology functions as a form of conflicted ecological imaginary, three of the six principles of heterotopia that Foucault outlines in 'Of Other Spaces' are particularly pertinent. Firstly, I consider Foucault's fourth principle: 'Heterotopias are most often linked to slices in time ... The heterotopia begins to function at full capacity when men arrive at a sort of absolute break with their traditional time'.[23] My next zooheterotopic principle emerges from Foucault's fifth, that 'heterotopias always presuppose a system of opening or closing that both isolates them and makes them penetrable'.[24] Lastly, I turn to Foucault's final principle: that heterotopias 'have a function in relation to all the space that remains'.[25] In order to illustrate the operation of these principles, I examine two definitive cryptid fictions that form a central part of the early twentieth-century proliferation of extra-zoological creatures crucial to the more serious academic work of Heuvelmans and others: firstly (and perhaps unavoidably) Conan Doyle's *The Lost World*; secondly, *King Kong*, both in Edgar Wallace and Merian C. Cooper's 1933 film and, more thoroughly, Delos W. Lovelace novelization (which unusually was released before the movie in 1932).[26]

Principle I: Heterochronies, or the Lands that Time Forgot

As much as cryptozoology is a discourse of space, it is also, connectedly, a discourse of time; or rather, more specifically, it is a particular sense of temporality that constitutes the spatiality, and vice versa. A location's geographical distance from the metropolitan centres of global capitalism allows for the possibility of an alternative relationship to global history. These are, to allude again to Edgar Rice Burroughs' famed story, the lands that time forgot: spaces that evade progress and modernity to offer a glimpse back into an ecology before the expansion of capital and taxonomy across the planet. Characteristically, in response to the spatial anxieties identified by Haggard and others, the cryptozoological adventure takes its heroes off the map and into a surviving area of blank space. The object of Professor Challenger's interest in *The Lost World* is an 'entirely uncharted' region of the 'partially explored' Amazon (p. 28). Likewise, as the crew of the 'moving picture ship' anxiously ponder their destination in Lovelace's neglected novelization of *King Kong*, the Hollywood filmmaker Carl Denham comments 'You won't find that island on any chart'.[27] The mysterious interior and the undiscovered island are perhaps the two most common spatial motifs of cryptozoological fiction, but there are other more flamboyant geographical premises too in related texts. Jules Verne's *Journey to the Centre of the Earth* (1864) as well as *The Land that Time Forgot* favour the epic literary device of katabasis, as the heroes descend into subterranean worlds that take them away from the world's mapped surface.[28] Such concealment routinely constructs a certain a-historicity; these are places where history has failed to happen.

The obsession with dinosaurs in cryptozoology and cryptid fictions serves as the exemplary sign of this breaking with chronology, of the stepping out of history into an era characterized, consistently, as the prehistoric. *King Kong*, not a text of course concerned primarily with dinosaurs, throws a handful in for good measure as the paradigmatic emblems of the space of the long before; as Lovelace describes, as a sign of the 'true scope of the island's mystery'.[29] 'Why shouldn't such an out-of-the-way spot', he asks, 'be just the place to find a solitary surviving prehistoric freak?',[30] a point reiterated throughout the text as Kong is understood immediately as 'prehistoric life' and a 'primitive survival'.[31] It is worth recalling that in a standard gesture of imperial fictions, human populations are also routinely located on a temporal axis; the sound of drums that greets the adventurers arrival on the mysterious Skull Island appear as 'the sound primitive hands might beat out in order to make easier for primitive minds the hard business of thinking'.[32] The film's racial suppositions are, of course, well known; the half-naked islanders (some in war paint, some dressed as gorillas) are blunt stereotypes of a generic cultural backwardness and the initial visual signature of the pre-modern world. Where the people are primitive, the logic runs, the wildlife will be too.

The Lost World goes even further back into a supposed human evolutionary past with the strange 'ape-men' that beset Professor Challenger and his crew on their South American plateau. Malone's first encounter with one of these unsettling beasts sees him linger over the Neanderthal features with an anthropologist's eye:

> It was a human face – or at least it was far more human than any monkey's that I have ever seen. It was long, whitish, and blotched with pimples, the nose flattened, and the lower jaw projecting, with a bristle of coarse whiskers round the chin. The eyes, which were under thick and heavy brows, were bestial and ferocious, and as it opened its mouth to snarl what sounded like a curse at me I observed that it had curved, sharp canine teeth[33]

Conan Doyle carefully constructs the ape as the proto-human, its face and voice just beneath what could be recognized as *Homo sapiens*, the 'thick and heavy brows' capturing the popular image of emerging man. The movement through space from the metropolitan centre to the colonial periphery that formulaically characterizes the opening chapters of cryptozoological fictions is very obviously, then, conceived as a movement into the past.[34] The representation of primitive cultures precisely captures Foucault's sense of the temporal dimensions of heterotopia; the functioning of heterotopia as heterochrony: 'as if the entire history of humanity reaching back to its origin were accessible in a sort of immediate knowledge'.[35]

Evidently, the motif of the prehistoric enclave serves a clear ideological function in the context of late imperialism, and one that is too widely recognized to be worth recounting here. Reading these texts ecologically, however, as texts concerned (however fantastically) with animal life does produce a degree of tension. For all the privileging of the modern West against its barbarian others, the *hauntique* and the landscape it inhabits is routinely expressed as an object of desire. There can be no more enticing lure to adventure than the prospect of a mythical beast. This lure, of course, has many aspects. Challenger's imperative to visit the Jurassic monsters in their 'secret haunts' in South America is framed as an attempt to recover his academic reputation in the face of the cryptozoological staple of a doubtful establishment. His companions embody a portfolio of other motivations; for Malone, it is the chance to win the heart of the impossible-to-please Gladys with her demand for 'a man of great deeds and strange experiences'.[36] The aristocratic adventurer, Lord John Roxton, on the other hand, is drawn to the prospect of the 'great wastelands and the wide spaces' of the Amazon.[37]

Following Roxton's ecologically resonant perspective, the text's most prominent initial construction of desire is based in an environmental aesthetics that Malone's narration draws consistent attention to, particularly in the first days of their South American sojourn. How, Malone asks as he surveys the glorious Amazonian vista, 'shall I ever forget the solemn mystery of it?' What follows is a deliberately (even stereotypically) romanticized evocation of the colonial landscape that insists on the aesthetic value of the world out of (Western) time:

> The height of the trees and the thickness of the boles exceeded anything which I in my town-bred life could have imagined, shooting upwards in magnificent columns until, at an enormous distance above our heads, we could dimly discern the spot where they threw out their side-branches into Gothic upward curves which coalesced to form one great matted roof of verdure, through which only an occasional golden ray of sunshine shot downwards to trace a thin dazzling line of light amidst the majestic obscurity. As we walked noiselessly amid the thick, soft carpet of decaying vegetation the hush fell upon our souls which comes upon us in the twilight of the Abbey, and even Professor Challenger's full-chested notes sank into a whisper.[38]

The reverent tone of Conan Doyle's writing revolves to a large extent around a sense of environmental nostalgia. Malone's recollection of his 'town-bred life' is the key structural point in the purple prose that follows; the aesthetic appreciation of the South American wilderness appears in contrast to the restricted urban world the novel begins in. The movement away from the metropolitan centre to the colonial wilds is constructed as a spiritual experience; 'even' the unsentimental Challenger is not immune to the hush that falls on the adventurers' souls. Entering the world before industrial development therefore involves an environmentalist (or proto-environmentalist) sensibility that identifies the pre-colonial wilderness as the home of a deep affective experience. Animal life features as a central ingredient of this landscape aesthetics as part of the mourned fullness of primordial nature. 'Once', Malone records, 'some bandy-legged, lurching creature, an ant-eater or bear scuttled clumsily amid the shadows'.[39] The indeterminate beast completes the scene, the uncertainty of its designation adding to the ecological ambience of a land that both exceeds and precedes the human; its scuttling movement among the shadows recalling the cryptozoological fixation with what cannot be clearly seen. Here, the world, exhausted by modernity, seems to flourish again.

If one key note of these cryptozoological fictions is the discovery of a nature that lifts the adventurer's heart, another equally recurrent dimension, somewhat paradoxically, is the presence of horror, an experience that extends beyond the straightforward physical terror that inheres in encounters with large, aggressive predators. It is not just the threat of the cryptid, but the *idea* of them that alarms the adventurer. For all the heroes' desire to uncover the plateau's lost creatures, as soon as they do, they wish they hadn't. Prehistoric life, above all, is slimy and disquieting, most notably in *The Lost World* when Challenger and his team meet with a 'rookery of pterodactyls':

> There were hundreds of them congregated within view. All the bottom area round the water-edge was alive with their young ones, and with hideous mothers brooding upon their leathery, yellowish eggs. From this crawling flapping mass of obscene reptilian life came the shocking clamour which filled the air and the mephitic, horrible, musty odour which turned us sick. But above, perched each upon its own stone, tall, gray, and withered, more like dead and dried specimens than actual living creatures, sat the horrible males, absolutely motionless save for the rolling of their red eyes or an occasional snap of their rat-trap beaks as a dragon-fly went past them.[40]

Malone's evocation of reptilian obscenity evidently constructs a wholly different atmosphere from his earlier experience of this zooheterotopia; his visceral reaction as he sickens in the company of these living fossils craves for the sanitizing sweep of imperial modernity to fight off the ontological vertigo the pterodactyls produce. Such 'filthy creatures'[41] demand violence and, sure enough, shortly afterwards the forest resounds with the reassuringly modern 'crash of Lord John's elephant gun' and a pterodactyl is discovered with a broken wing 'spitting and gurgling ... with a wide-opened beak and blood-shot, goggled eyes, like some devil in a medieval picture'.[42] The entry into prehistory is ultimately a compromised experience, as if there is only so far back you can go before modernity (initially bemoaned) entices you back. Prehistoric life (the embodiment of what we might term zooheterochrony) is a site of both environmental nostalgia and profound ontological discomfort, and as such, of a profound ambivalence.[43]

Principle II: Opening/Closing; Isolation/Penetration

The interplay between desire and disgust in *The Lost World* illustrates the significance of cusps and transitions between contrasting zones of experience, an aspect of zooheterotopias that is recurrently condensed into the image of the portal or gate. Evidently, as Foucault explains his fifth principle, 'the heterotopic site is not freely accessible like a public space'.[44] As we have seen, it is the quality of being somehow cordoned off, removed or restricted that constitutes the (zoo) heterotopia. Foucault's description of the 'system of opening and closing that both isolates [heterotopias] and makes them penetrable' is pertinent to cryptozoology not just because of the geographical inaccessibility of the cryptid's home, but also because of a kind of ritual that appears to govern the entry of the cryptozoological adventurer into these set-apart places. Through these, zooheterotopias, as much as they associate capitalist modernity with imaginative and aesthetic loss, are also embroiled in a masculinist identity politics inseparable from imperial ideologies.

To continue with *The Lost World*, at the outset of his heroes' adventure, Conan Doyle constructs an elaborate narrative structure of obstruction and penetration that insists on the closure of the secret domain, but which in doing so provides the clue to its opening. As Professor Challenger explains as the men approach the plateau:

> the secret opening is half a mile onwards upon the other side of the river. There is no break in the trees. That is the wonder and mystery of it. There where you see light-green rushes instead of dark-green undergrowth, there between the great cotton woods, that is my private gate into the unknown. Push through and you will understand.[45]

The concealed gap in the undergrowth (the first of several portals the heroes have to negotiate in the novel) is a well-established literary device; the same basic premise, indeed, as the wardrobe in C. S. Lewis's Narnia stories from the 1950s. Clearly, the primary function of such portals is to emphasize the destination as a place beyond and constitute the world's 'lostness'. These are the geographical curiosities that allow rare and fantastic species to flourish, a premise that makes sound ecological sense. As Darwin found in the Galapagos, geographical isolation is of profound ecological significance in the evolution of distinct evolutionary forms. Immediately, animal life beyond Challenger's 'private gate' is more plentiful: monkeys, tapirs and a puma appear along a stretch of water in which '[b]ird life was abundant'.[46] Whatever is 'unknown' is automatically understood as ecologically flourishing, as if the act of knowing is itself a violent act that reduces biodiversity. Most significantly, the 'deep peace of this strange waterway was unbroken by any sign of man'.[47] The gap in the undergrowth is the portal between the human world and the more-than-human world beyond. Importantly, access is restricted to a privileged few; as Foucault suggests, 'To get in one must have a certain permission and make certain gestures'.[48] The image of a 'private gate' recalls Challenger's wariness of visitors to his house earlier in the novel (resulting in Malone's unceremonious eviction from the premises) and indicates the importance of exclusion as an aspect of concealment, a motif that forms a central part of the imperialist dimensions of such texts. Even before a formal incursion into the territory, Challenger's remark lays down a marker of sovereignty. The gate (and implicitly the land beyond) is his own domain, beyond the reach of other visitors.

Importantly, not everyone will be proved fit to traverse the gate and enter the interior. This restriction produces a particular narrative experience of the threshold that evidently provides a good deal of the text's early excitement as the dauntless, hardy explorer ritualistically proves himself by fighting his way in. *King Kong*'s gate into the unknown is altogether a clearer obstacle than that of *The Lost World*. Skull Island is protected by the 'mighty barrier' of a wall in which 'there seemed to be a gate hinged to massive stone pillars' that Denham and his crew can only reach by traversing a 'tortuous passageway'.[49] Such a proliferation of impediments creates the impression that the island's interior will be 'impregnable',[50] though inevitably the story's narrative drive is structured around the dramatic penetration of this barrier. Crossing the portal in *King Kong* is given an extra ideological dimension by the kidnap of the blonde starlet Ann Darrow (Fay Wray) that demands the adventurers' intervention to save her from a dire fate at the hands (paws?) of the beast-God Kong. *King Kong* is noteworthy for the games it plays with the extension of desire across categories of difference (species most obviously, but also, implicitly, race); but what is important in terms of its operation as a zooheterotopia is the containment of the act of territorial penetration within a masculinist and heteronormative context. Consequently, while the cryptozoological fantasy of habitats out-of-time articulates a resistance to the effects of imperial capital, the trope of the threshold reveals an ideological and aesthetic investment in what you might think of as its causes: a dominatory sensibility that also, of course, demands the destruction of the monsters on the threshold's other side.

After recounting Kong's retreat back into the island's interior with his terrified human prize, Lovelace turns in a new chapter to the frantic quest to retrieve her:

> It had been Denham who raced the rescue boats away from the Wanderer; and Denham who had deployed the sailors for the breathless run to the village. But from the moment the great gate swung open Driscoll took charge. It was Driscoll who organised the pursuit after Kong.[51]

As the director and first mate set off on their desperate mission, Lovelace insists on the military decisiveness of the two men as the sailors are 'deployed' and the pursuit 'organised'. Shortly afterwards, Driscoll and Denham are described inspecting 'guns, ammunition and flashlights'[52] as their penetration of the zooheterotopic threshold is structured around stock images of masculine accomplishment. Importantly, the gate effects a sudden shift in the hierarchical structure of the pursuit as Driscoll immediately takes charge, transformed from first mate to leader on the other side of the threshold. The centrality of the love interest between Ann and Driscoll in *King Kong* (in contrast to its marginality in *The Lost World*) reaffirms the rather Darwinian stakes of the enterprise: kill the beast and get the girl.[53] The zooheterotopia, therefore, becomes less a discourse of the other and more a discourse of the self in the face of the other. Nostalgia for the lost world of the *hauntique* is the occasion for the triumph of modernity over the past through the determination of its reach into the world's furthest corners. The function of the portal as the site of heroic self-assertion (a trope also apparent in the latter stages of Conan Doyle's narrative) adds a strikingly normative element to the cryptozoological fixation with the beyond. That which is set-apart from capitalist modernity is useful for the opportunity it provides to showcase the macho prowess of the adventurous hero as the lost habitat appears as a space in which to reconstruct a pioneer sensibility ostensibly fading in the modern world. What this second function of the zooheterotopia confirms, then, is a well-remarked convergence of what may appear to be an environmentalist aesthetic with the history of empire.[54]

Principle III: Zooheterotopia/Globalization

The function that Foucault concludes with, that 'heterotopias have a function in relation to all the space that remains' evokes a state of hyper-connectedness characteristic of what today would most commonly be identified as globalization. In part, the zooheterotopia fits with Foucault's discussion the heterotopia of 'compensation',[55] specifically in the way that the cryptid's abundant and mysterious world responds to the imaginative and ecological lack of industrial modernity. But the function of relation characteristic of *The Lost World* and *King Kong* is determined more prominently and consistently by an economic imperative to render the cryptid visible. The zooheterotopia cannot remain secluded, its threshold must be traversed and, most importantly, its denizens must be returned to the imperial metropolis in order to be fully integrated into the capitalist world system. The cryptid must be brought 'home'. For all the formulaic inevitability of this narrative pattern, there is, nonetheless, a degree of uncertainty to the way in which the cryptid's return operates as a sign of incipient globalization with all the ecological consequences that carries. Ultimately, it is precisely the texts' representation of the cryptid's resistance to commodification that illustrates what is most important about cryptozoology's contribution to an ecological imaginary in the era of global capital; namely, the way that ecological thinking can be incorporated (albeit uneasily) into global capital.

As been pointed out by a number of critics, the adventures of both *The Lost World* and *King Kong* are framed by economic discourses. In *The Lost World*, the introduction of capital is rather subtle, or at least more subtle than in *King Kong*. The novel begins with a knowingly dry reflection on 'bimetallism' as the father of Malone's beloved Gladys embarks on a 'monotonous chirrup about bad money driving out good, the token value of silver, the depreciation of the rupee, and the true standards of exchange'.[56] Such humdrum reflections are evidently gauged to be in stark contrast to the excitement that follows. They are significant nonetheless for their foregrounding of an economic context. As Ross. G. Forman has argued, the framing conversation about markets is inextricably connected to the novel's intertextual relations: *The Lost World* 'follows adventure conventions by encoding crucial economic allegories in fictional discourse'.[57] Moreover, the novel ends with the image of a 'beautiful glittering diamond' extracted from the secret plateau that invites the adventurers back to South America to claim their full rewards as imperial entrepreneurs. Once discovered, the cryptozoological habitat must be instrumentally utilized.

The economic agenda of *King Kong* is even more immediately striking. Kong exists from the start to be monetarized in the developing movie industry that the text is metafictionally based around. Set in the midst of the great depression, the story begins in grinding poverty with the down-on-her-luck starlet-to-be Darrow caught stealing an apple and redeemed from prosecution from the director Denham who pays off the shop proprietor. The quest for Kong is always concerned, then, with questions of livelihood. Accordingly, an economic dimension of the quest for Kong is signalled repeatedly in the narrative's early stages. The theatrical agent Weston emerges from a cab 'hanging tightly to his money'[58] and the question of how much the movie will gross is a central part of planning discussions.[59] Most obvious is Denham's over-excited exclamation in the film after Darrow is rescued from the beast: 'we've found something worth more than all the movies in the world'. Kong, as Lovelace has a tabloid photographer remark, is a 'gold mine'.[60]

Both texts, therefore, require us to understand the hunt for the cryptid as fundamentally engaged with an economic logic.[61] Before the adventurers' destinations are reached, Skull Island and Conan Doyle's South American plateau are already understood to be connected to all other sites through the profit motive, though it would be an oversimplification to think of either text as straightforwardly complicit with the advance of global capital. As Barbara Creed observes 'New York is represented as an uncanny modern urban jungle in which greed is the ruling principle'.[62] Lovelace's novel provides notable evidence of the irony that surrounds Kong's final commodification in a New York theatre, trussed up for the audience's edification. If Kong has disappeared into consumer culture, then so have the gathering throng looking for their hit of the spectacular. Waiting for the curtain to reveal the sensational gorilla, the crowd are reduced to metonym: a '$3.98 pick-me-up frock' converses with a 'Bronx Derby' while a 'Paris Gown' condemns them as a 'rabble' to the chagrin of a 'Riverside Drive tip-brim'.[63] Lovelace's exposure of class inequalities that emerge in a society governed by consumption clearly adds a political edge to the drive to total commodification that the quest for Kong appears to represent. The narrative's setting against the background of the great depression does, moreover, remind us of one of the great, periodic failures of capitalism.

A certain discomfort with economic drives is apparent in the story's poignant conclusion. Kong's escape from the theatre is based not just on his saccharine love for Darrow, but also on his immediate and visceral resistance to the waves of flash photography that greet his unveiling. There is perhaps a slight biblical resonance to Lovelace's presentation of this climactic moment. Twice the 'blinding glare' of the photographers' cameras fills the stage before Kong's rage brings the situation to crisis point:

> The white glare flashed across the stage a third time. Kong opened his mouth and roared from deep in his chest. As Driscoll swung an arm protectingly around Ann the captive beast-god struggled furiously. His rage was a wild and cataclysmic emotion which surged from his inmost being.[64]

Lovelace's insistence on the three flashes emphasizes Kong's role as a sacrificial, even (at the risk of monstrous over-reading) a Messianic figure, recalling Peter's denial of Christ three times, before Kong is finally laid low clinging to the Empire State Building, that unmistakable emblem of capitalist bravado.[65] Reading Kong as ape-Christ is perhaps going too far, but his rebellion and then fall from grace clearly overthrow his potential as a 'gold mine'; the narrative experience of his defeat revolves around the tragic sense of his 'inmost being' that refuses the apparent logic of a total commodification. Kong, the great crypto-zoological icon is, finally, more than his image. Consequently, it is the relation of Skull Island to all other sites that is the most fundamental and ambivalent aspect of its zooheterotopic function. The demand for the economic use of a being whose value is constituted by his exteriority to economic use results in the paradox of wanting something and not wanting it at the same time.

The Lost World presents a rather different resolution to the cryptid's 'return' to the metropolis. Here, a pterodactyl is brought back to London to be exhibited before an incredulous public in the heart of the commercial city. Its inevitable escape, 'squeezing its hideous bulk' through an open window represents the novel's narrative climax, though its fate is ultimately more ambiguous than Kong's.

'Nothing can be said to be certain upon this point', Malone concludes of the monster's fate, before giving one final, vague suggestion:

> The only other evidence which I can adduce is from the log of the SS. Friesland, a Dutch-American liner, which asserts that at nine next morning, Start Point being at the time ten miles upon their starboard quarter, they were passed by something between a flying goat and a monstrous bat, which was heading at a prodigious pace south and west. If its homing instinct led it upon the right line, there can be no doubt that somewhere out in the wastes of the Atlantic the last European pterodactyl found its end.[66]

For all the novel's determination to incorporate the lost plateau into imperial capital, like Kong, the forlornly homing pterodactyl is a poignant figure. As Forman argues, '*The Lost World* converts the monster into the imperial victim, not the victimizer'.[67] Conan Doyle's reinsertion of the pterodactyl into the epistemological lack that allows cryptozoology's sense of possibility, leaves us back where we started. Who knows if the monster isn't lingering somewhere in the 'wastes of the Atlantic'? The economic drive that brings the zooheterotopia into contact with all other sites must always, it seems, be incomplete; the effects of capital always appear in suspension or under denial.

Both these texts, then, and cryptozoology more broadly, serve to imaginatively construct the limits of capital in relation to animals and their dwindling habitats. Ironically, it is the imagination of the limits of capital that appears to drive its forward momentum. Foucault's enigmatic final sentence in 'Of Other Spaces' offers a parting reflection on the way that heterotopias are involved with the imagination when he turns to ships as heterotopias *par excellence*, and the 'greatest reserve of the imagination'. 'In civilisations without boats', Foucault contends, 'dreams dry up, espionage takes the place of adventure and the police take the place of pirates'.[68] Foucault signals a spatial imaginary as a counter hegemonic force; to be out of the way is to be located at a remove from normative ideology in a space that allows us to reimagine the world and, potentially, our relations with other creatures. This might seem to allow for an optimistic conclusion to the role of the zooheterotopic imagination in revivifying ecological energies in an era of crisis. Any optimism must be necessarily cautious, however. *The Lost World* and *King Kong* ultimately revolve around the capitalist logic of scarcity: whatever is lost, singular, secret or unobtainable reaches the status of a kind of super-commodity. To follow Conan Doyle's logic: when a monster is lost, it must be found again; once it is found, it must be lost again, if only to raise the prospect that it might once more be found. What this leaves us with is the sense of the emotional and political complexity of the cryptozoological habitat. The very act of bemoaning the effects of capitalist modernity becomes in turn another *part* of capitalist modernity as part of a fantasy of infinitely continuing possibility at odds with the material grounding of finite planetary life.

10 SOFT MACHINES

Fabienne Collignon

Michel Foucault's 'Of Other Spaces' posits the existence of 'counter-sites', which he defines as 'kind of effectively enacted utopia in which the real sites ... that can be found within culture ... are simultaneously represented, contested, and inverted'.[1] These 'heterotopias' occur 'by way of contrast' to utopias,[2] alternate systems of being that are, as Fredric Jameson argues, radically other, by which he means that the form itself – what he also refers to as the utopian 'program' – reflects on difference in a late consumer capitalist state where every other possibility always already seems exhausted: the future will have failed.[3] Jameson's utopian argument is, according to Phillip E. Wegner, a utopian 'problematic', interrogating reality through a contested idea, that is, utopia, considered as a dialectic, impossible and yet indispensable, because trying to imagine a 'not yet' (as) opposed to the present order:[4] it stands against the 'invincible universality of capitalism'[5] at the same time that it attests to capitalism's catastrophic power. The 'best Utopias', according to Jameson, 'are those that fail the most comprehensively' because as records of 'ideological imprisonment',[6] they might nonetheless yield the means to neutralize the present and future-as-insolvent. He understands utopia, then, as potentially transformative; the dynamics of utopian politics is its dialectic of 'Identity and Difference, to the degree to which such a politics aims at imagining, and sometimes even ... realizing, a system radically different from this one'.[7] It can be a space of revolutionary practice, invariably closed or seeking closure, read autonomy, because removed: a zone apart, a new spatiality committed to (a precarious) totality, with us here and them over there, beyond the moat, the wall. He also distinguishes, however, a utopian 'impulse' that he defines as a 'specialized hermeneutic of interpretive method', encompassing political practice, even 'liberal reforms and commercial pipedreams, the deceptive yet tempting swindles of the here and now'.[8] He recognizes a spatial distinction between this program, that is, dreams of enclosure, and an impulse as a hermeneutic, but what the utopian vocation really means is the ability or compulsion to think a break. In this way, and considering that 'in the absence of reliable content only form can fit the bill' – because the utopian content can only ever work in terms of a 'critical negativity' without being able to offer up any viable alternatives – the form itself is rupture, the 'radical closure of a system of difference in time', in which the aftermath is secondary: the vital political function of the utopian form is to 'force' a breakage.[9]

'[F]undamentally unreal spaces',[10] as Foucault describes them, he is less concerned with the radical potential, or even spirit, utopias might have. He considers them either perfections or else inversions – as 'inverted' analogies of 'the real', they imagine 'society turned upside down' – whose production is negligible compared to 'other sites' that 'contradict' the present order of being.[11] Elsewhere/nowhere, their spatiality, existing purely as textuality, does not interest Foucault here, in this manuscript, 'Of Other Spaces', that, unrevised or not reviewed for publication (it was initially included in the October 1984 edition of the French journal *Architecture–Movement–Continuité*), is itself oddly situated with respect to the 'official corpus'[12] of Foucault's work. In the spring 1986 issue of *Diacritics*, in which Foucault's article is first published in English, a footnote comments on the space of the text above, which 'retains the quality of lecture notes',[13] exhibited in Berlin shortly before his death: it is itself 'a contestation of the space in which we live'[14] and cease to live. The essay, though unequivocally declaring to be about space at the expense of time or history, nonetheless remains concerned with losing time, the 'loss of life', dissolution, disappearance.[15] From the very start – and bearing in mind the footnote-underworld that refers to the circumstances of the article's publication – there is a sense of time running out: heterotopias are linked to time through structures that accumulate it, in museums, libraries, cemeteries, or that exist in transitory terms, as festivals, fairgrounds, these usually 'marvellous empty sites on the outskirts of cities'.[16] Yet even the archive, as Jacques Derrida argues, is preoccupied with time and with its own ruin, which constitutes its condition of existence;[17] even in apparently 'eternal' places, where time 'never stops building up' and which behave as if 'outside of time and inaccessible to its ravages',[18] it is the possibility of 'remainderless destruction'[19] that governs their state of being. It is, consequently, difficult to interpret heterotopias purely in terms of an 'epoch of space', which, at any rate, has a place in a 'history of space',[20] because the prospect of death/extermination so clearly informs their emergence. Foucault's opening statement, particularly, resonates with an investigation that seeks to interrogate the wind turbine and wind farm as both symbols of utopian thought and marked out or marketed 'counter-sites' to entropy-inducing fossil fuel and nuclear technologies.

'Of Other Spaces' begins with the following observation:

> The great obsession of the nineteenth century was, as we know, history: with its themes of development and of suspension, of crisis and cycle, themes of the ever-accumulating past, with its great preponderance of dead men and the menacing glaciation

of the world. The nineteenth century found its essential mythological resources in the second principle of thermodynamics.[21]

Reading the text today, it is, in the first instance, glaciation that is menaced, the past and future that are threatened, the theme of development – the fetishization of interminable growth – invariably associated with the second principle of thermodynamics. If, then, wind farms function as a heterotopic counter-force to the law of entropy, their 'enacted utopia', at the outer edges of cities, like fairgrounds, proposes to 'accumulate time' through a renewable energy process in which resources 'won't run out'.[22] They are, as such, 'oriented toward the eternal',[23] but their status as heterotopias is absolutely contestable, particularly in relation to colonies of 'clean' machines, whose presence, after all – in spatial border zones, at short distances from urban centres, in moorland, for example, that has been 'improved' or 'restored' according to 'habitat management plans'[24] – is due to large multinational companies. The means by which these farms suspect or invert the prevailing systems of energy production is, as a result, especially problematic: if anything, the small percentage of renewables, at once marginal and so vital to marketing strategies, resolutely affirms the dominance of these conglomerates which don't, of course, desire any change and instead seek to safeguard the eternity of their interests – they are colonial outposts in the service of an empire that hopes to never end. And yet, and this is no doubt partly the outcome of successful promotional campaigns, colony and turbine still function in terms of a utopian imagination (a utopian text occupies reality too; its form, perhaps, itself a heterotopia.) Real spaces, wind farms nonetheless also remain fundamentally unreal: they exist, but are at the same time otherworldly, as if their presence here on earth was only an indication of things to come, of an actual, radical disruption that has not yet arrived.

Contradictory sites, wind farms consequently stage a crisis of temporality too, at present, in the present order, a manifestation of corporate megawattage that in no way incapacitates the empire and simultaneously – because and despite of the promotion machine – a zone of futurity, reaching beyond the here and now towards a utopian life generated by way of clean energy. To look at, for example, the online presence of Whitelee Wind Farm, situated not far from Glasgow and the largest onshore installation in the UK, is to notice the predominance of kids; their visitor centre offers interactive exhibitions, recreational activities, an educational programme about energy consumption and renewable energy plans. The future is at stake here, and the narrative ostensibly is one of corporate, and nuclear (family), responsibility that begins today, at both local/individual and global level, though the latter really only operates as window-dressing, the 'swindles' that Jameson mentions and that continue to function as 'lure and bait' for capitalist ideology, to which the child similarly belongs.[25] But as much part of a political practice that in actuality dreams of immobility – business as usual – these sites might retain their function as a promise to inaugurate, 'on the other side of the glass',[26] as it were, a new epoch of 'future development', not least because at Whitelee, on Eaglesham Moor, visitors can enter into communication with monumental machines. The managed environment here, still suitably 'wild', if mainly due to the weather and wind – the moor is a flatland where vantage points reach far, to outlying islands – is at once 'naturally' and technologically sublime: this space/time configuration constitutes a curious zone of pilgrimage, hostage to the conception of a never-ending empire and in thrall to what could be, a coming, harmonious techno-futurity whose advent is the turbine.

More likely than not, this 'future development' is the result of a 'benign' capitalism, where corporations behave 'ethically' – there is always a limit, evidently, not least because this type of 'ethics' still purely operates for profit – yet the possibility remains that this colony '[gives] over to the infinity of the sea',[27] that is, to an imagination that, aware of limits, might no longer be bound by them. This is, in fact, Jameson's argument, too, implied by the closing lines of Foucault's article; the problem that poses itself, as ever, is how exactly to breach these limits, bearing in mind that even in the most generous of analyses, a wind farm, under corporate control, never actually constitutes, but only mimics, a rupture. In this vein, the following essay stages an investigation that while not wanting to dismiss the potential for some kind of departure – a piece of space as boat-like in its disposition – nonetheless stays vigilant regarding the (maybe largely mythical) production of this spatiality as heterotopia. The chapter further focuses on the singular turbine as concrete expression of a utopian impulse that predominantly emerges due to the machine's sublimity: the turbine's thrust demands a techno-fetishist attention, so joined up, as it is, with the utopian form as culmination of technological achievement and aesthetics.

Mirrored Stages

Foucault's discussion on the concept of heterotopia as 'counter-site' includes the example of a mirror; between 'real places' and utopia emerges a 'mixed' space – like the looking glass as both place and placeless – that 'crosshatches'[28] lines of division. Foucault writes:

> In the mirror, I see myself where I am not, in an unreal, virtual space that opens up behind the surface; I am over there, there where I am not, a sort of shadow that gives my own visibility to myself, that enables me to see myself there where I am absent: such is the utopia of the mirror. But it is also a heterotopia in so far as the mirror does exist in reality, where it exerts a sort of counteraction on the position that I occupy.[29]

The subject location is split in two, though the split self remains off-limits: the mirror's interweaving of spaces is the focus here, as a surface that, yielding worlds beyond its framework, is both 'absolutely real' and 'absolutely unreal'.[30] It is this passage and the material reality/unreality of the mirror that serve as point of entry into an interpretation of the wind farm as a heterotopic site, not only because it functions, as mentioned earlier, as a zone that is at once totally present, in the present, over here, and at the same time curiously other: a device that has arrived from the future, a placeless place, not yet represented or fully 'in reality'. It is, however, also the materiality of the turbine that reflects this 'sort of counter-action'[31] to the position of a presence or present: the machine is made of (fibre) glass, a synthetic material that, mirror-like, extends the limits of possibility. To expand, then, past the reach of the present, into a realm that 'opens up behind the surface', towards an absence from which the current order of being can be reformed: wind farms, if considered beyond the political/economic conditions of their production, establish oppositional zones that, to a certain extent, interrogate the dominant system of extractive technologies. Over here, however, on this side of the mirror, these 'farms' – a word that comes with such deep connections to the land – are subsumed into larger networks of energy generation that stay carbon-centred, though they remain, either way, a visible indication of crisis, a reminder of fossil fuel scarcity and, perhaps, the harbingers of an apocalypse.

They are, to be sure, science fictional (SF) devices, future forms that 'stalk' the horizons of SF films: J. J. Abrams's *Mission Impossible III* (2006), in which a field of turbines stands against a deep blue sky, in twilight, itself indicative of their intermediate, shadow occupation – half in this world, half out of it. It seems, at first, a space of serenity, the sublime play of light and darkness returning a suggestion of quietude that assumes a position of contrast to the abandoned warehouse somewhere in Berlin that serves as the temporary base of operations for terrorists. Effectively, though, the constellation of turbines becomes a maze-like realm that the helicopters of both government and rogue agents must navigate: this field, after all, is coldly neutral, indifferent, a sublime, complex area, whose sense of quiet order is only apparent from the outside, not the inside, which is all chaos, movement, slice.[32] It is even more remarkable that the film's plot centres around the recovery of a 'dark' technology (as enigmatic as the 'Red Matter' in Abrams's 2009 *Star Trek*) that Benji's (Simon Pegg) Oxford Professor refers to as the 'Anti-God', an 'accelerated mutator like an unstoppable force of destructive power'.[33] The world, according to this man, 'would eventually be eviscerated by technology', a predictive statement that foresees the evisceration of secret agents, recruited to preserve order; whether this evisceration occurs by way of brain-bomb implant, turbine or 'Anti-God' ultimately matters little – all technology, each of which exists as another manifestation of an unknown, therefore 'dark', science or matter, is invariably neutralized by the good Americans. The shot, turbines in twilight, is indicative of this machine existing not only with reference to overtly disastrous technology but demonstrates that its object ultimately is a Janus-faced engine, whose bi-faced nature also recurs in Christopher Nolan's *Inception* (2010).

Inception, in opposition to extraction (to steal dream-ideas), means origination, hence gestation; the film, in many ways, is about planting – planting a 'foreign' idea in the mind of Robert Fischer (Cillian Murphy; dreaming subject), inheritor of his father's energy company that if remaining intact, as empire, according to one of his competitors, functions like a 'new superpower'.[34] Inception should appear as organic, the process of entering and implantation disguised as self-originating: an appearance of 'natural' growth as cover-up for a penetrating, potentially deadly, violence. To plant an idea in the heir is to break up superpower dominance, a rationale that itself obfuscates the film's justification for what remains an act of taking by force and colonizing process; the suggestion, at the end of the film, is that to split Fischer Morrow into component parts also signals a departure from the reliance on extractive technologies, because the implanted idea (empire break-up, connected with an ice-cold father figure, to instead make the son 'grow' something of his own) is conceptualized as pinwheel windmill, functioning as a fake symbol of innocence and paternal love. At Whitelee Wind Farm, workshops teach kids to make their own pinwheel windmills; in the film, windmill/turbine, though linked to the dismantling of an energy conglomerate, nonetheless remains a device that signals an assault, masking its occurrence through a supposedly 'natural' process of origin and growth, further yielding false memories of the Father (and never mind the central blind spot of *Inception*, that language is already implantation). Empire is counter-acted by a reversal of the colonizing force, directed against its own subject and establishment, but this strategy, at any rate not without its victims, aligns alternatives to extraction and superpower with a deceptive pure-heartedness, the myth of the natural, while the law of the Father remains fundamentally unchallenged: the windmill in the safe of the inheriting mind itself inherits this myth.

If the turbine, as SF device, operates as convincing symbol of revolt, as in Frederik Pohl and Cyril M. Kornbluth's 1952 novel *The Space Merchants* – set on a ravaged earth, where politics really is pure advertising – it is clearly in the hands of a revolutionary force, in this case in exile on Venus, which eventually functions as a place of safety for insurgents, its atmospheric energy tapped through turbines. These modes of production, distinguished through direct control and access, as well as presumably entailing a responsible and reduced energy usage, cause an emancipation from an oppressive regime and thereby provide the means of sustenance for an alternative way of being in (and out of) this world. David Dickson, in a study published a year after the 1973 oil embargo, discusses such types of non-invasive technologies, whose designation shifts, reflecting the discrepancies inside the 'alternative technology movement'; they are variously called 'soft technology, radical technology, low impact technology, intermediate technology (applied to technological requirements of under-developed countries), people's technology, liberating technology'.³⁵ The names are variations on a theme: soft machines propose to stand in opposition to tyrannical power and to prevent the emergence of an ice-planet and/or threat of entropy that Foucault mentions at the start of his article. Considering, then, that this techno-infrastructure – identified through its adjectives, soft, low, democratic – aims to reduce environmental impact, the exploitation of individuals, and to advocate regional self-sufficiency, it could, as a result, mark out a heterotopic counter-site, to allow a life on 'the other side of the glass'³⁶ and 'beyond the end of history'.³⁷

Yet to split the enquiry into two parts, separating a liberating from an enslaving technology – totally recalling the T-101 in *The Terminator* (1984) in contrast to *T2* (1991), in which Arnold Schwarzenegger's killing machine/father figure becomes plain Mutilator: the Terminator no longer kills in the sequel; he only wounds – indicates the fallacy of technological neutrality that Marshall McLuhan examines in *Understanding Media* (1964). McLuhan argues against 'the voice of current somnambulism', which is predicated on precisely this myth of technology as fundamentally neutral:

> Suppose we were to say, 'Apple pie is in itself neither good nor bad; it is the way it is used that determines its value'. Or, 'The smallpox virus is in itself neither good nor bad; it is the way it is used that determines its value.' Again, 'Firearms are in themselves neither good nor bad; it is the way they are used that determines their value.' That is, if the slugs reach the right people firearms are good. If the TV tube fires the right ammunition at the right people it is good.³⁸

McLuhan's point, here, is that the nature of the medium is left unexamined, as are the processes by which new technologies are developed or promoted; what happens instead is a blank dismissal of the politics of killing machines/power sources depending on who uses or programs them. This splitting up creates binary conflicts: good/bad, renewable/non-renewable, soft/hard, democratic/despotic technology. The forward slash implies a radical break, a mark that dismisses the continuities between these apparent opposites; it is an easy way out for those that seek to uphold the status quo. Following Slavoj Žižek – who noted that we can now have 'coffee without caffeine, cream without fat, beer without alcohol' – Mirko Zardini writes that,

> [w]hile we can accept a temporary disruption for a short period (such as the 1973 oil crisis) as a way to restructure our productive system, our desire is always to maintain our current lifestyle, economic system, and social organisation, and to fuel them with alternative sources of energy (possibly green) while reducing the polluting effects ... In this sense, a zero-emission, zero-energy building would fit neatly between coffee without caffeine and war without warfare.³⁹

The initiatives to simply replace the means by which energy is obtained – from extraction to 'inception', that is, to the beginning of relations that apparently contradict or dismantle the empire – do not, then, usually engage in any sustained effort to interrogate either the distribution of power or, beyond that, the dominant ideology of consumer capitalism. Hence the divisions between good and bad machines: the preservation of an order depends on that distinction, the false logic of a somnambulist. The issue of the wind farm as heterotopia consequently becomes difficult, if not impossible to defend; the turbine for the most part forms another device that legitimizes the established regime, even if the potential of resistance remains – off-planet, for example, in a future/*avenir* not yet here, over there, beyond.

To return, as such, to the mirror – 'outside of all places'[40] but still present – I would, however, like to suggest another process by which to analyse a turbine, the single entity that generally/conceptually stays a part of a corporate totality. Roland Barthes' *Mythologies* (1957), notably the entry on the Eiffel Tower, becomes a point of reference in this case, interested in turbine mythology accessed through an attention to the material character of the engine, whose affinities lie less with the mirror than with related forms and vectors of light. If Lewis Mumford writes that the paleotechnic age, dominated by coal, is swathed in this substance's colours, 'grey, dirty, brown, black' which 'spread everywhere',[41] then the renewable energy industry aligns itself with – note the emergence of yet another (false) binary – illumination. Never mind, for now, the provenance of glass, transmuted from ashes into transparency – 'darkness', after all, as Don DeLillo observes, 'is just another name for light'[42] – it is this alliance with mirror-like conditions, and therefore with a heterotopic site, that warrants consideration. The focus on turbine tower and revolving blades at their tips is at once techno-fetishist and oppositional: it is a study of things in parts, in close proximity to technics. By combining techno-fetishism and Luddism, gadget lover and machine breaker, the emergent property will be one that crosses those easy distinctions between renewable and non-renewable energy sources to reveal the extent to which wind turbine technology still carries onward the 'great itineraries' of our mechanical dreams.[43]

The point of convergence is, then, the medium – the blades more so than the tower, though both form, of course, an integrated structure; a gadget lover arises through the glance into a mirror or pool of water, as McLuhan argues: 'Narcissus mistook his own reflection in the water for another person. This extension of himself by mirror numbed his perceptions until he became the servomechanism of his own extended or repeated image.'[44] At the base of the Narcissus myth is gadget love, the subject merged with a machine; in this vein, the medium as extension of the flesh involves a 'state of numbness' that McLuhan also calls 'Narcissus-narcosis'.[45] As such, in the 'true Narcissus style of one hypnotized by the amputation and extension of [my] own being in a new technical form', to look at the medium, in a way and to a point – so as not to encompass 'total field-awareness'[46] – already means to be enthralled. The medium in the first instance refers to the material out of which it is made, and the ideologies that both form and substance carry: an investigation of a turbine in light of machine aesthetics and machine affiliations, about steel (tower) and plastic (blades), about reflections that, according to Fernand Lèger, '[fill] the role of unlimited fantasy'.[47] A tower stands – dreams of escape are propelled onwards by a stationary vehicle –, its 'simple, primary shape confers upon it the vocation of an infinite cipher'.[48] To further appropriate Barthes, whose sentence runs on in an acknowledgment of infinity, the tower, read turbine, also prompts a list of other 'ascensional dream[s]', the 'rocket, stem, derrick, phallus, lightning rod or insect'.[49] If the substitution of Eiffel Tower with turbine seems facile, the intention of this exchange is to primarily highlight those correspondences obscured by forward slashes. The renewable/non-renewable binary is not treated explicitly, but approached tangentially via a series of other seemingly antithetical forces; the objective is to establish continuity, to incorporate the turbine into a narrative of 'ascensional dreams', whose thrust stays that of the (war) machine and, above all, of the rocket.

Turbine, Rocket, Stem, Derrick, Phallus, Lightning Rod, Insect

If Barthes' essay functions as a guideline for a possible cultural and mythological study of a turbine, then those aspects that are pushed to the margins by the prominence of the good killing machine – that is, the accumulation of factors that include the false logic of a technological neutrality which is, to a certain extent, disclosed in Barthes' enumeration above – have to be hauled into the centre of analysis. Often interpreted as bi-faced, as mentioned earlier, technology is suspended between orders – the good or evil Twin – and transcends, as Barthes perceives the Eiffel Tower, the realm of reason and clear function. The Tower is dream-like, an 'infinite cipher', which Barthes senses as a 'friendly' presence;[50] if this mythology can be applied to a wind turbine, whose symbolic force at times outweighs its functional operations, what should follow is a similar concern with affect. Whatever the Tower might signify in the future, Barthes is convinced that 'it will always be something ... of humanity itself';[51] 'friendly', it is also of this earth. Yet he misses an important connection here, although his list of comparisons hints at what is left out, namely the Tower or technology's connections to science fiction. He mentions the rocket, so clearly an SF device, and moves on, from there, to include giant organisms, plants and insects that appear gigantic because compared to colossal mechanisms like space ships and oil rigs, or also gantries. To overlook the sentiment that such devices, concurrently biological and mechanical, exude – bear in mind, for example, H.G. Wells' *The War of the Worlds* (1898), in which the invading Martians are at once fungoid and technicized[52] – is to ignore the estrangement caused by machines which SF, as a literature of rupture,[53] brings to light. Barthes' chain of associations, which should have lead him to conclude differently, or at least to allow for some doubt to destabilize the foundations of his hyperbole, is blind to the catastrophic possibilities of chain reactions.

Let's be receptive to these orders of existence, consequently, and admit that turbines do not necessarily and exclusively speak of 'humanity': monolithic, monochromatic, they might have arrived from 'outside', announcing the war of the worlds.[54] Turned into the wind, they are strangely sentient, ready to meet, to support or oppose an invasion: they might be with us, or against us. After all, the rhetoric of sustainability is war-like, to the point that distinctions between military and environmental issues are close to vanishing points.[55] In the US, as Adrian Parr observes, the threat to national security now encompasses environmental degradation; in 1990, Al Gore published a document titled 'Strategic Environment Initiative' (SEI), a reference to Ronald Reagan's Strategic Defence Initiative (SDI), whose degrees of fictitiousness are encapsulated in its popular designation, Star Wars.[56] Gore's report called for the development of environmentally friendly technologies for energy, transportation, manufacture, but at the same time, his document proposed a realignment of soft technology with military hardware. (One wonders why Gore, however ironically, or no matter how pragmatic in his appeal to a core Republican audience largely hostile to green politics, would choose to hitch his proposal to such an obvious fantasy, though the myth of Ronnie Raygun in the meantime might well overshadow SDI's technological impossibility). At the risk of repetition: turbine and rocket, particularly in terms of SDI rhetoric and as long as it is not in the service of an 'other' – recall *Mission Impossible III* – are 'friendly' gadgets whose primary purpose is defence, against the 'menacing glaciation of the world'.

The correlations between phallic turbine and war machine are not all purely aesthetic, but are also evident in terms of a conceptual engagement, that is, the sustainability enterprise so easily incorporated into the martial discourse. This might be the reason why Timothy Morton's difficult book *Ecology Without Nature* (2007) abounds with SF or Cold War rhetoric and mechanisms – he begins his study by referring to the domino theory and keeps using words that belong to the vocabulary of the rocket: escape velocity, gravitational fields, orbit. The effort to sustain the planet is framed with reference to Cold War culture, which, despite Space Age gadgets, stays committed to policies of containment. A mirror image of the rocket, it cannot simply be said that the turbine – which like the space ship, also at first a vehicle of escape, is about limits, the terminal points of resources on a finite planet – forms part of a whole other set of principles that is established according to dividing lines: the one as the other's evil Twin. They are both indicative of a potential global meltdown and of a world under siege whose glaciation must, at the same time, both be prevented and preserved – a world at a standstill, watched over by the rocket/turbine, keeping things as they are; this reference extends to the poles, which soft machines protect like guardian angels. Rocket and turbine mass can, as already intimated before, be read according to similar parameters that move past their phallic symbolism; their glossy skins draw attention to surface politics.

Fascist-Fetishism

To look at turbines according to their materiality brings to mind modernity, twentieth-century matter like plastic, shimmering textures, the fascist-futurists (Marinetti's romance with brutal materialism, for instance), the works of Charles Sheeler, American painter (1883–1965), whose classical landscapes monumentalize the machine age, concurrently seductive and sinister. Functionality, as Lèger argues, is beautiful, but coldly so: the smooth-edged fairings and streamlining techniques of a machine aesthetics throw dark shadows although it is, of course, light that is at stake here. In his 1914 essay 'Geometric and Mechanical Splendour and the Numerical Sensibility', Marinetti describes the technological sublime, whose 'essential elements' are: 'hygienic forgetfulness, hope, desire, controlled force, speed, light, willpower, order, discipline, method ... the concurrence of energies as they converge into a single victorious trajectory'.[57] The 'victorious trajectory' of these attributes or energies which Lèger, in 1925, likewise celebrates – he stands in awe in front of the 'manufactured object' that he perceives as 'a polychrome absolute, clean and precise'[58] – is that of a supreme dominance of the technological entity and, further, of its totalizing systems. What lies buried in such 'energies', in words such as hygiene and cleanliness, is whiteness, fascism: the state of beauty and 'truth', Lèger's expression for absolute utility (e.g. beauty), is achieved through a eugenic/streamline procedure divesting objects or bodies of 'parasite drags'[59] that obstruct movement. Christina Codgell argues that eugenics and streamline design emerge at the same time, equally considered to be 'agents of reform' and promoting evolutionary progress; both, she says, were obsessed with increasing efficiency and hygiene, in order to create a 'perfect' specimen as the means to achieve an imminent 'civilised utopia'.[60]

If the prevailing aesthetic of the 1930s was a design of pure surface, Charles Sheeler's paintings – smooth, flawless, curiously without depth – exemplify this form or 'movement', though all is, in Sheeler's work, still and silent: his perspectives present beauty as a void. Objects stand in an eternal vacuum, apparent on various levels, an emptiness of presence as well as an absence of motion. Sheeler's *Suspended Power* forms part of a series of paintings commissioned by *Fortune* magazine in 1938; published as a portfolio on the theme of 'Power', the painting, as the title already indicates, captures the nature of this power as just that, suspended – not simply physically, hanging, as it does, at the point of falling, but also metaphorically, as a force whose temperament is not quite clear. The blade cuts an atomized figure in half; the rest of the bulk, typically accomplished in such a manner so as to erase evidence of the artist's execution, suggests a ballistic missile whose titanic warhead is freeze-framed just before it hits ground zero: classical landscapes of power arrest a world right before its meltdown and disappearance in light.

Although Edmund Burke, in *A Philosophical Enquiry into the Origins of Our Ideas of the Sublime and Beautiful* (1757), wrote that the sublime is mostly associated with darkness, which is 'more productive' to call forth feelings of sublimity, he concedes that 'extreme light' is 'by its very excess ... converted into a species of darkness'.[61] It is, however, not the surplus but the play of light that Léger comments on at the *Salon d'Automne* in Paris one year: the 'beautiful, metallic objects, hard, permanent, and useful, in pure local colours; infinite varieties of steel surfaces at play next to vermilions and blues. The power of geometric forms dominated it all'.[62] The machine object, displacing light, is sublime but the attention lavished on such surfaces gestures towards a psychological condition or sexual aberration, as Freud notes in his article on 'Fetishism'. There exist elements of endlessly delayed gratification with regards to the technological sublime, suggesting progress *ad infinitum* but always tinged with the possibility of extinction – the threat of annihilation hangs suspended – which, held back, is pleasurable.

Freud discusses his 'opportunity of studying', over the course of years, 'a number of men whose object-choice was dominated by a fetish', which emerged tangentially in analyses designed to treat other abnormalities.[63] One of his case studies – the 'most extraordinary' situation – concerns 'a young man' who 'had exalted a certain sort of "shine on the nose" into a fetishistic precondition'.[64] The man, notes Freud, had been brought up in an English nursery, but then moved to Germany, where he had almost forgotten his mother-tongue. 'The fetish', Freud proceeds, 'had to be understood in English, not German. The "shine on the nose" [in German "*Glanz auf der Nase*"] – was in reality a "*glance* at the nose". The nose was thus the fetish, which, incidentally, he endowed at will with the luminous shine which was not perceptible to others'.[65] Freud claims, and this comes as no surprise, that the fetish is a substitute for 'a particular and quite special' penis, namely the mother's, and that it serves as a 'token of triumph over the threat of castration and a protection against it'.[66] If, however, the fetish is examined in German, as *Glanz*, it is tied to light, as well as to the phallus: the fetish as light-induced or light-related. Retaining Freud's interpretation of the fetish's bifurcated significance of both horror and salvation, it triumphs over a threat which is, yet again, suspended: it functions as 'the last impression before the uncanny and traumatic'[67] realization of the mother's lack, which always stands at the verge of being revealed. A 'counter-wish'[68] to the finality of exposure and therefore violence, the 'shine' of the fetish easily catches the gleaming skin of metallic object-desires.

Dreams of Plastic, Dreams of Flow

To recall Barthes' chain of associations and return to the SF-tech of phallic war technology (bear with me; turbines approach the horizon): in Thomas Pynchon's *Gravity's Rainbow* (1973), Margareta Erdmann works as a movie actress in 'dozens of vaguely pornographic horror' films during the twenties and thirties.[69] Erdmann, whose modified name is 'politically safe' in Nazi Germany, a reference to the fascist 'code' of Earth, Soil, *Volk*,[70] is involved in what she calls a séance with a group of men including SS *Obersturmführer* Dominus Blicero, her captor and game keeper, and IG Farben *Generaldirektor* Smaragd. They all progress into a board room, where the men sit around a conference table, in the middle of which lies something 'gray, plastic, shining, light moving on its surfaces'; Erdmann is then dragged off into a warehouse, past 'great curtains of styrene or vinyl, in all colours, opaque and transparent, hung row after row from overhead. They flared like the northern lights'.[71] She is stretched out on an inflatable plastic mattress:

> Someone said 'butadiene', and I heard *beauty dying* ... Plastic rustled and snapped around us, closing us in, in ghost white. They took away my clothes and dressed me in an exotic costume of some black polymer, very tight at the waist, open at the crotch. It felt alive on me. 'Forget leather, forget satin', shivered Drohne. 'This is Imipolex, the material of the future'.[72]

Imipolex G is a 'new plastic, an aromatic heterocyclic polymer'; in conjunction with steel, plastic forms such a 'delight to the fetishist'.[73] Imipolex G is used for a special rocket's insulation device, a polymer that wraps around a white boy, gagged with a white glove – the 'symbolism is all engineered' by Blicero – and inserted into the modified rocket designed to leave the planet 'in [sexual] love: so taken that ... death, and life, will be gathered, inseparable, into the radiance' of the take-off.[74] It is not, however, only the rocket that combines these substances: wind turbines are similarly composed of steel and plastic in an arrangement that is beset with paradoxes. To name but a few: their blades or wings touch the sky; they manifest a dream of function; a soft technology, they form part of a rhetoric and aesthetic of a (white-washed) future. Aerodynamic considerations shape the design; it is easy to link them to rocket and gantry, the blades to propellers – with reference, again, to *Mission Impossible III* – and to point out shared definitions and developments of mechanical parts like pods, or nacelles. They belong, then, as much to earth than to the air or outer space, their rotations so clearly are part of the myth of technological revolution: fantasies of mechanical ascent beyond the dying planet. Static, they nonetheless embody dreams of flight; the tension between immobility and propulsive thrust adds to those other ostensibly antithetical properties subsequently united into one form, such as the requirement for both lightness and strength. The steel tower, which burrows into deep intraterrestrial space – it rests on foundations which weigh around 750 tons – is outfitted with fibreglass blades, a lightweight synthetic material that allows them to bend into the wind. The lines are smooth, streamlined; the circular motions of the propeller-blades perform long sweeps, the graceful movements of a *ballet mécanique*.

The point is that by way of their material – and irrespective, for now, of their usage, which makes all the difference if it occurs in correspondence to a change of heart, or politics, to programs of disruption – turbines continue to propagate a very particular discourse that is founded on the idea of the machine as salvation. Machines by design are obsolescent, to be replaced by other devices that might reflect contemporary concerns or efforts to reduce carbon emissions. In and of themselves, turbines are only devised to uphold the existing state of affairs; their entrenchment in this conservative politics is further compounded by a fetishist-analytical interpretation of fibreglass and, therefore, of plastics. V. E. Yarsley, and E. G. Couzens' 1941 book, simply titled *Plastics*, features an airplane on the cover, a white plastic bird called the Timm Trainer, a fitting image to announce the beginning of a 'Plastic Age': an epoch to take to the air. The authors proceed to give a brief historical survey in which they note that 'from 1925 onwards [,] the success of these new products was evident' and that 'the public' has since become 'plastics conscious'.[75] They cite the mass production of the phenolic and amino-plastics as the reason, yet they are blind to and blinded by the mythology of plastics, which forms the continuity of the narrative. Right from the start, they write that a 'plastic material is ... one which at some stage in its history was capable of flow, and which on the application of the necessary heat and pressure can be caused to flow and take up a desired shape'.[76] The 'desired shape' is achieved through the exploitation of molecular possibilities, which explode the limits of nature and power into infinity. In an extraordinary passage in the last chapter titled 'Plastics and the Future', Yarsley and Couzens speak about the 'thousand and one uses of plastic', the stories of total control and total flexibility existing in an environment of high polish made of 'light, smooth, non-corrodible' substances.[77] The material combines 'great strength with low specific gravity' flowing into a form of 'gently curving' lines, a structure that is rivetless and minimizes friction: ultimate strength lies in tension, which 'ensures that there is no creep'.[78]

Turbine blade, made of fibreglass and wood, has to be placed in context, into a semiotics of plastic;[79] the material's development coincides with streamlining design evocative of speed and the 'desire', according to Jeffrey Meikle, 'for frictionless flight into a utopian future whose rounded vehicles, machines and architecture would provide a visually uncomplicated protective environment ... closed off from the Depression and marked by static perfection'.[80] As Meikle develops elsewhere, the ascendancy of plastics did not generally happen until the late 1920s and early 1930s – think of those other ventures that wish to leave the planet and find expression in, for example, Fritz Lang's *Frau im Mond* (1929), a film about rocket flight – even though the patent on celluloid, the oldest plastic, was taken out by Alexander Parkes in 1855.[81] The 'Plastic Age' is not proclaimed until later however; industry insiders began to announce it in 1927 and in the following years dreams of technological utopia – evidenced as much by statements about plastic possibilities than by the formation of Rocket Societies working to achieve a functional escape vehicle[82] – really took off. In his book *Form and Re-Form* (1930), Paul T. Frankl, an American architect and furniture designer, writes that the new materials 'speak in the vernacular of the 20th Century', and articulate 'the language of invention [and] synthesis', claiming that,

> We are no longer preoccupied with our past. We are piercing the future. Not merely are we looking into a future – we are actually being propelled into a future: a morrow more thrilling, more breathtaking, more compelling in its dictatorship than the flimsy traditions of the 19th Century.[83]

Frankl echoes the Futurists, even more so in *Machine-Made Leisure* (1932), where he announces that 'We love [the machine]' in 'breathless adoration',[84] in the rapture of the techno-sublime. He holds out the promise of a future that is configured in terms of the constructions of modernity: the skyscraper, the department store, the airplane, but the device that lurks in the shadows, and is developed under the auspices of a German dictatorship, is the rocket.

Plastic 'consciousness' expanded in the 1930s, the decade when 'materia nova' – Frankl's term – no longer simply fake nature but 'bring to the service of industry an array of new properties – the new gifts', according to John Gloag, 'of lightness, translucency, transparency, texture and colour'.[85] While he remains ambiguous about the substances and transforming technologies, he is nonetheless seduced by their properties and possibilities, the aesthetic brilliance that is exhibited at World Fairs in New York and San Francisco in 1939. Like Henry Adams, in the great gallery of machines at the Paris exhibition in 1900, Gloag, confronted with the 'remarkable ideas about the alliance of light with transparent and translucent materials', comments on 'the enclosing of objects in irregular masses of transparent plastic, so that they interrupted and distributed a beam of tinted light'.[86] They perform, then, tricks of light, allow to transmit it in an 'infinity of variations' for the purpose of blinded consumption; what further helped to swell public awareness of the new matter was the development of thermo plastics, forming 'a galaxy of new materials'[87] encompassing, for example, polystyrene and polyethelene. The definition of plastic, though – 'applied to anything which possesses plasticity, that is, anything which can be deformed under mechanical stress without losing its cohesion, and is able to keep the new form given to it'[88] – also applies to thermo setting plastics, most common in the interwar period. These are, once subjected to heat and pressure, converted into 'insoluble, infusible masses, which cannot be further re-formed':[89] they are stable, inert, immortal, and retain the shape of their first, and only, moulding. Thermo plastics, on the other hand, capable of being softened and re-softened indefinitely, suggest, as Roland Barthes writes, 'the very idea of ... infinite transformation', so that plastic is 'less a thing than the trace of a movement',[90] a means by which to 'torch the past'[91] and boost a culture of impermanence through the mobility of molecules.

Fibreglass is a glass-reinforced polymer, a material whose plastic matrix can be both thermo-setting and thermo-plastic. From the very start of its invention, it had no clear identity – it is known by many names (its name is Legion) but lacks, so Meikle writes, the high-tech prestige of nylon or Mylar.[92] In *Gravity's Rainbow*, Pynchon refers to the very tight relationship between bombs and plastic, both developed at Du Pont; the connections between synthetics and death – early newspaper articles on nylon reported that the material was made out of human corpses pulverized into 'cadaverine'[93] – take place on multiple levels, from source to development concurrent with, and applications in, weapons systems. Mylar, similarly hitched to rocketry, reaches outer space as the inflatable balloon-shaped Echo I satellite, a communications device launched in 1960 which functioned as a reflector, not transmitter, of telephone, radio and TV signals.[94] Yet nonetheless, fibreglass –developed during World War II as a replacement to plywood in order to shield aircraft radomes and also used for communication purposes, including in water, gas, and sewage systems – contains a reference to techno-dreamworlds by way of the Monsanto House of the Future, developed at MIT in the 1950s under the direction of Albert Dietz. The building – whose 'total enclosure is defined by a "continuous surface"' of glass-reinforced polyester[95] – was completed in 1957 and exhibited at Disneyland, where else, where it stood, for a period of ten years, as a celebration of plastic possibility. By 1967, it had come to embody failed potential; despite the promised flexibility of the material, the House – to which in theory levels could be removed and added, re-arranged at will – was in fact permanent, fixed in time with respect to its streamlined past but also regarding the difficulty with which it was finally dismantled. Both House and, by extension, the substance that realized it store and relay impressions of an impossible future, one of extraordinary technological promise that does not come to pass or passes too quickly, revealing the fundamental flaws of its design. Like so many technologies, fibreglass went ballistic – linked, obviously or insidiously, to military operations (Monsanto manufactured the rainbow herbicide Agent Orange) – not least in terms of its return to earth: after its flight of fancy, it must crash back down.

Wind turbine blades retain a plastic legacy, the myth of a soft technology that paradoxically desired, in some of its applications, to avoid leaving enduring damage but carrying out just that: with its million-year survival rate, plastic forms new continents in the Pacific. Turbines, evidently, are invested with the belief in a wasteless or waste-diminished world: they are dismantled easily, unlike the Monsanto House of the Future, but like Echo I, they tend to merely deflect the rays of investigations. Incorporated into an agenda that amalgamates environmentalism, capitalism and militarism, they stay passive devices, still carrying the high sheen of technological euphoria that allows the continuation of business/war by other means. The narrative of extraction – from a carbon, entropy-inducing economy, however tenuous this process still is – has to be accompanied by a reflection on 'needs', such as the 'need' for an unlimited supply of energy on a limited planet, functioning as part of the familiar rhetoric which shifts as new discoveries are made: more electricity to produce more plastic wings. In his book *The Technological Society* (1964), Jacques Ellul asks: 'Doubtless, technique has its limits, but when it has reached these limits, will anything exist outside them?'[96] Ellul raises a pertinent question about a closed world where gadgets fit into existing networks, the circulatory apparatus of a regime where, to return to Mirko Zardini's comment, renewable energy sources 'fit in neatly between coffee without caffeine and war without warfare' and where words like freedom, as Esther Leslie concluded at a conference at Birkbeck in May 2011, transform into military manoeuvres: Operation Freedom. Rather than 'spaces of alternate ordering',[97] which is how Kevin Hetherington understands Foucault's term, giant wind farms, as corporate zones, adhere to, or feed, a system of rapacious energy consumption; under such circumstances, they fail to challenge the 'need' upon which the network, Vampire/Empire Electric, thrives. As such, in this system, wind turbine aggregations are no indication of a heterotopic counter-site to the 'invincible universality of capitalism' but tend to maintain technological somnambulism: to offer an alternative way of being in the world, opening up on the other side of their plastic coating, they would have to announce total rupture with the present/presence rather than functioning as a program that seamlessly continues the practices of the 'eternal' empire. It is, after all, as if their mirrored surfaces, here, project no reflection over there; the 'absolutely unreal space' that Foucault attributes to an 'other side', at once absent and present, is fully lacking in this instance, these corporate fields, in which soft machines keep rotating, without causing revolutions, in the service of precisely this horizon-less order: the future as already neutralized in advance.

NOTES

Introduction

1. E. Soja, *Thirdspace: Journeys to Los Angeles and Other Real-and-Imagined Places* (Oxford: Wiley-Blackwell, 1996), p. 162.
2. P. Johnson, 'The Geographies of Heterotopia', *Geography Compass*, 7:11 (2013), pp. 790–803, on p. 790.
3. H. Heynen, 'Afterthoughts: Heterotopia Unfolded?', in M. Dehaene and L. De Cauter (eds), *Heterotopia and the City: Public Space in a Postcivil Society* (London: Routledge, 2008), pp. 311–24, on p. 312.
4. R. J. Topinka, 'Foucault, Borges, Heterotopia: Producing Knowledge in Other Spaces', *Foucault Studies*, 9 (2010), pp. 54–70, on p. 55.
5. Johnson, 'Geographies of Heterotopia', pp. 796–7.
6. J. D. Faubian, 'Heterotopia: An Ecology', in M. Dehaene and L. De Cauter (eds), *Heterotopia and the City: Public Space in a Postcivil Society* (London: Routledge, 2008), pp. 31–40, on p. 31.
7. M. Foucault, 'Of Other Spaces', trans. J. Miskowiec, *Diacritics*, 16:1 (Spring 1986), pp. 22–7, on p. 27.
8. Johnson, 'Geographies of Heterotopia', p. 790.
9. Soja, *Thirdspace*, p. 159.
10. M. Dehaene and L. De Cauter, 'Heterotopia in a Postcivil Society', in M. Dehaene and L. De Cauter (eds), *Heterotopia and the City: Public Space in a Postcivil Society* (London: Routledge, 2008), pp. 3–10, on p. 4.
11. 'Heterotopic', Merriam Webster Dictionary, at http://www.merriam-webster.com/dictionary/heterotopic [accessed 24 August 2014].
12. A. Saldanha, 'Heterotopia and Structuralism', *Environment and Planning*, 40:9 (2008), pp. 2080–96, on p. 2080.
13. Cited in M. Foucault, *The Order of Things* (New York: Vintage, 1970), p. xv.
14. Foucault, *The Order of Things*, p. xviii.
15. For a full account of the development and various transitions of Foucault's essay, see P. Johnson, 'Unravelling Foucault's "Different Spaces"', *History of the Human Sciences*, 19:4 (2006), pp. 75–90. An exhaustive set of resources on Foucault and heterotopia is also available on Johnson's website: P. Johnson, *Heterotopia Studies: Michel Foucault's Ideas on Heterotopia*, at http://www.heterotopiastudies.com.
16. Foucault, 'Of Other Spaces', p. 22.
17. Foucault, 'Of Other Spaces', p. 22.

18. Foucault, 'Of Other Spaces', p. 23.
19. B. Latour, 'Spheres and Networks: Two Ways to Reinterpret Globalization', *Harvard Design Magazine*, 30 (2009), pp. 138–44, on p. 144.
20. M. Foucault, 'The Eye of Power', *Power/Knowledge: Selected Interviews and Other Writings, 1972–77* (New York: Pantheon, 1980), p. 149.
21. N. Thrift, 'Overcome by Space: Reworking Foucault', in S. Elden and J. W. Crampton (eds), *Space, Knowledge and Power: Foucault and Geography* (Basingstoke: Ashgate, 2007), pp. 53–8, on p. 55.
22. Foucault, 'Of Other Spaces', p. 24.
23. Foucault, 'Of Other Spaces', p. 24.
24. Foucault, 'Of Other Spaces', p. 24.
25. Foucault, 'Of Other Spaces', p. 25.
26. Foucault, 'Of Other Spaces', p. 25.
27. Foucault, 'Of Other Spaces', pp. 25–6.
28. Foucault, 'Of Other Spaces', p. 26.
29. Foucault, 'Of Other Spaces', p. 26.
30. N. Fraser, 'From Discipline to Flexibilization? Rereading Foucault in the Shadow of Globalization', *Constellations*, 10:2 (2003), pp. 160–71, on p. 166.
31. J. Read, 'A Genealogy of Homo-Economicus: Neoliberalism and the Production of Subjectivity', *Foucault Studies*, 6 (2009), pp. 25–36, on p. 33.
32. Foucault, 'Of Other Spaces', p. 22.
33. Latour, 'Spheres and Networks', p. 141.
34. Latour, 'Spheres and Networks', p. 144.
35. Johnson, 'Unravelling Foucault's "Different Spaces"', p. 87.
36. See Dehaene and De Cauter's editorial material in note 15 of M. Foucault, 'Of Other Spaces', trans. M. Dehaene and L. De Cauter, in M. Dehaene and L. De Cauter (eds), *Heterotopia and the City: Public Space in a Postcivil Society* (London: Routledge, 2008), pp. 13–29, on p. 25.
37. Johnson, 'Unravelling Foucault's "Different Spaces"', p. 84.
38. M. Warner, 'Diary', *London Review of Books*, 36:17 (2014), pp. 42–3, on p. 42.
39. Wagner, 'Diary', *London Review of Books*, p. 43.
40. P. Rabinow, *The Foucault Reader* (New York: Pantheon Books, 1984), p. 252.

1 From Hegemony to Heterotopias: Geography as Epistemology in Gramsci and Foucault

1. M. Foucault, 'Some Questions from Michel Foucault to Herodote', in J. W. Crampton and S. Helden (eds), *Space, Knowledge and Power: Foucault and Geography* (Burlington: Ashgate, 2007), pp. 19–20, on p. 19.
2. J. B. Racine and C. Raffestin, *Response*, in J. W. Crampton and S. Helden (eds), *Space, Knowledge and Power: Foucault and Geography* (Burlington: Ashgate, 2007), pp. 31–3, on p.31.
3. Foucault, 'Some Questions from Michel Foucault to Herodote', p. 21.
4. Foucault, 'Some Questions from Michel Foucault to Herodote', p. 21.
5. S. Elden and J. W. Crampton, 'Introduction', in J. W. Crampton and S. Helden (eds), *Space, Knowledge and Power: Foucault and Geography* (Burlington: Ashgate, 2007), pp. 1–16, on p. 1. See also M. Tanca, *Geografia e filosofia* (Milan: Franco Angeli, 2012).

6. C. Philo, 'Foucault's Geography', in M. Crang and N. Thrift (eds), *Thinking Space* (London: Routledge, 2000), pp. 205–38, on p. 221.
7. M. Foucault, *The Archaeology of Knowledge* (New York: Pantheon, 1972), p. 7.
8. M. Sheridan, *Michael Foucault: The Will to Truth* (London: Tavistock, 1980), p. 49.
9. M. Foucault, *Language, Countermemory, Practice* (Ithaca, NY: Cornell University Press, 1977), p. 139.
10. S. Vaccaro, 'Introduzione', in M. Foucault, *Spazi altri. I luoghi delle eterotopie* (Milan: Mimesis, 2008), pp. 7–16, on p. 8.
11. G. Deleuze and C. Clement, 'Dall'Anti Edipo a Mille Piani', in G. Deleuze (ed.), *Felicità nel divenire. Nomadismo, una vita* (Milano: Mimesis, 1996), pp. 41–51, on p. 48.
12. M. Foucault, *Power/Knowledge: Selected Interviews & Other Writings* (New York: Pantheon, 1980), p. 68.
13. Philo, 'Foucault's Geography', p. 208.
14. M. Foucault, 'Truth and Power', in P. Rabinow (ed.), *The Foucault Reader* (Harmondsworth: Penguin, 1984), pp. 51–75, on p. 59.
15. Foucault, *The Archaeology of Knowledge*, p. 10.
16. Foucault, *The Archaeology of Knowledge*, pp. 8–9.
17. C. Minca, *Introduzione alla geografia postmoderna* (Padua: CEDAM, 2001), p. 47.
18. M. Foucault, 'Of Other Spaces', *Diacritics*, 16:1 (1986), pp. 22–7.
19. M. Serres, 'The Geometry of the Incommunicable: Madness', in A. Davidson (ed.), *Foucault and his Interlocutors* (Chicago, IL: Chicago University Press, 1997), pp. 36–56, on p. 48.
20. F. Driver, 'Geography and Power: The Work of Michel Foucault', in P. Burke (ed.), *Critical Essays on Michel Foucault* (Cambridge: Cambridge University Press, 1992), pp. 148–68, on p.150.
21. Foucault, *Power/Knowledge*.
22. Philo, 'Foucault's Geography', p. 229.
23. F. Frosini, *Gramsci e la filosofia* (Rome: Carocci, 2003), p. 15.
24. A. Gramsci, *Quaderni del carcere*, ed. V. Gerratana (Turin: Einaudi, 1975), p. 856.
25. Gramsci, *Quaderni del carcere*, p. 1233.
26. A. Burgio, *Gramsci storico: Una lettura dei Quaderni del carcere* (Bari: Laterza, 2002), p. 3.
27. Frosini, *Gramsci e la filosofia*, p. 80.
28. A. Gramsci, *Prison Notebooks*, ed. J. A. Buttigieg, 2 vols (New York: Colombia University Press, 1996–2011), vol. 2, p. 195.
29. Frosini, *Gramsci e la filosofia*, p. 81.
30. G. Liguori, *Sentieri gramsciani* (Roma: Carocci, 2006), p. 169.
31. Burgio, *Gramsci storico*, p. 3.
32. A. Gramsci, *Lettere 1908–1926*, ed. A. Santucci (Einaudi: Turin, 1992), p. 130.
33. A. Gramsci, *La questione meridionale*, ed. F. De Felice and V. Parlato (Rome: Editori Riuniti, 2005), p. 176.
34. Gramsci, *Quaderni del carcere*, pp. 2021–2.
35. A. D. Morton, 'Traveling with Gramsci: The Spatiality of Passive Revolution', in M. Ekers, G. Hart, S. Kipfer and A. Loftus (eds), *Gramsci: Space, Nature, Politics* (Oxford: Wiley-Blackwell, 2013), pp. 47–64, on p. 48.
36. J. Schneider, *Italy's Southern Question: Orientalism in One Country* (Oxford: Berg, 1998).
37. Gramsci, *Quaderni del carcere*, p. 475.

38. Gramsci, *Quaderni del carcere*, p. 476.
39. Gramsci, *Quaderni del carcere*, p. 353.
40. F. De Felice and V. Parlato, 'Introduzione', in Gramsci, *La questione meridionale*, pp. 56–7.
41. G. Liguori, in F. Frosini and G. Liguori, *Le parole di Gramsci* (Rome: Carocci, 2004).
42. Gramsci, *Quaderni del carcere*, p. 703.
43. Gramsci, *Prison Notebooks*, vol. 2, p. 91.
44. Gramsci, *Quaderni del carcere*, p. 703.
45. Gramsci, *Quaderni del carcere*, p. 937.
46. E. Said, 'History, Literature, and Geography', *Reflections on Exile* (Cambridge, MA: Harvard University Press, 2002), p. 458.
47. J. Wainwright, 'On the Nature of Gramsci's Conceptions of the World', in M. Ekers, G. Hart, S. Kipfer and A. Loftus (eds), *Gramsci: Space, Nature, Politics* (Oxford: Wiley-Blackwell, 2013), pp. 161–77, on p. 167.
48. Gramsci, *Quaderni del carcere*, p. 1935.
49. Foucault, *Language, Countermemory, Practice*, p. 148.
50. Foucault, 'Of Other Spaces', p. 23.
51. Gramsci, *Quaderni del carcere*, p. 70.
52. Gramsci, *Quaderni del carcere*, p. 2141.
53. Gramsci, *Quaderni del carcere*, p. 72.
54. Foucault, 'Of Other Spaces', p. 24.
55. Gramsci, *Quaderni del carcere*, p. 2148.
56. G. Liguori and P. Voza, *Dizionario Gramsciano 1926–1937* (Rome: Carocci, 2009), p. 684.
57. Gramsci, *Quaderni del carcere*, p. 2149.
58. Gramsci, *Quaderni del carcere*, p. 2167.
59. Gramsci, *Quaderni del carcere*, p. 2163.
60. B. Han, *Foucault's Critical Project: Between the Transcendental and the Historical* (Stanford, CA: Stanford University Press, 2002), p. 116.
61. M. Foucault, *Discipline and Punish: The Birth of the Prison* (London: Penguin, 1991), p. 137.
62. M. Foucault, *The History of Sexuality* (New York: Vintage, 1990).
63. Gramsci, *Quaderni del carcere*, p. 2149.
64. C. Minca, *Introduzione alla geografia postmoderna* (Padua: CEDAM, 2001), p. 52.
65. Foucault, 'Of Other Spaces', p. 23.
66. A. Gramsci, *Quaderni del carcere*, ed. V. Gerratana (Turin: Einaudi, 1975), p. 857.
67. Gramsci, *Quaderni del carcere*, p. 1430.
68. N. Srivastava, *The Postcolonial Gramsci*, ed. N. Bhattacharya and B. Bhattacharya (London: Routledge, 2012), p. 2.
69. E. Said, *Reflections on Exile* (Cambridge, MA: Harvard University Press, 2002), pp. 464, 467.
70. G. Baratta, *Le rose e i quaderni. Il pensiero dialogico di Antonio Gramsci* (Roma: Carocci, 2003), p. 100.
71. F. Frosini, *La religione dell'uomo moderno. Politica e verità nei Quaderni del carcere di Antonio Gramsci* (Rome: Carocci, 2010), p. 35.
72. Frosini, *La religione dell'uomo moderno*, p. 124.
73. A. Loftus, 'Gramsci, Nature and the Philosophy of Praxis', in M. Ekers, G. Hart, S. Kipfer and A. Loftus (eds), *Gramsci: Space, Nature, Politics* (Oxford: Wiley-Blackwell,

2013), pp. 178–96, on p. 184.
74. A. Gramsci, *Selections from the Prison Notebooks*, ed. Q. Hoare and G. Nowell Smith (London: Lawrence and Wishart, 1971), p. 354.
75. Gramsci, *Quaderni del carcere*, p. 1437.
76. D. Boothman, 'Translation and Translatability', in P. Ives and R. Lacorte (eds), *Gramsci, Language and Translation* (Lanham: Lexington Books, 2010), pp. 107–33, on p. 108.
77. Gramsci, *Quaderni del carcere*, p. 1575.
78. D. Harvey, 'The Kantian Roots of Foucault's Dilemmas', in J. Crampton and S. Helden (eds), *Space, Knowledge and Power*, pp. 41–9, on p. 44.
79. Harvey, 'The Kantian Roots of Foucault's Dilemmas', p. 188.
80. C. Minca, *Introduzione alla geografia postmoderna* (Padua: CEDAM, 2001), p. 49.
81. E. Morera, 'Gramsci's Critical Modernity', in M. Green (ed.), *Rethinking Gramsci* (London: Routledge, 2011), pp. 238–65, on p. 244.

2 'An Occult Geometry of Capital': Heterotopia, History and Hypermodernism in Iain Sinclair's Literary Geography

1. G. Deleuze and F. Guatarri, *What is Philosophy?*, trans. G. Burchell and H. Tomlinson (London: Verso, 1994), p. 96.
2. M. Serres and B. Latour, *Conversations on Science, Culture and Time* (Ann Arbor, MI: University of Michigan Press, 1990), p. 60.
3. G. Poynter and I. MacRury, 'Preface', in G. Poynter and I. MacRury (eds), *Olympic Cities: 2012 and The Remaking of London* (London: Ashgate, 2009), pp. xiii–xvi, on p. xv.
4. Massey reminds us of the crucial role that London institutions played in conceiving and disseminating neo-liberal instruments, which have become hegemonic since the 1980s. D. Massey, 'Opportunities for a World City: Reflections on the Draft Economic Development and Regeneration Strategy for London', *City: Analysis of Urban Trends, Culture, Theory, Policy, Action*, 5:1 (2001), pp. 101–6; and D. Massey, *For Space* (London: Sage, 2005). See also M. Edwards, 'What Ever Happened to *Capital* in Working Capital?', *City: Analysis of Urban Trends, Culture, Theory, Policy, Action*, 10:2 (2006), pp. 197–204; and M. Edwards, 'Structures for Development: Getting Them Right', in P. Cohen and M. Rustin (eds), *London's Turning – Thames Gateway: Prospects and Legacies* (London: Ashgate, 2008), pp. 283–92.
5. R. Hewison, *The Heritage Industry* (London: Methuen, 1987); and C. Jencks, *The Language of Post-Modernist Architecture* (London: Academy Editions, 1987), p. 95.
6. A. Huyssen, *Present, Pasts, Urban Palimpsests and the Politics of Memory* (Stanford, CA: Stanford University Press, 2003), p. 7.
7. This tradition runs from G. White, *Natural History and Antiquities of Selbourne* (London: Printed by T. Bensley for B. White and Son, 1789) to W. G. Sebald, *Rings of Saturn* (London: Harvill Press, 1995). In his afterword to the Hackney photographer Stephen Gill's *Archaeology in Reverse*, Sinclair notes that Hackney photographer Gill's sensitive, local experiential vision is 'wary of that mendacious conceit, "closure"'; I. Sinclair, 'Diving in to Dirt', in S. Gill, *Archaeology in Reverse* (London: n. p., 2007). Sinclair has spoken of the need to exercise this aesthetic method driven by 'flow [and] momentum' to disclose 'accumulations' in two debates surrounding the 2012 Olympics. See I. Sinclair, Keynote speech given at 'Life Outside the Blue Fence', Planners Network UK (PNUK) and Games Monitor Conference, Limehouse Town Hall, London, 10–11

April 2008; and L. Porter, 'Reflections on the Event', Report on 'Life Outside the Blue Fence', Planners Network UK (PNUK) and Games Monitor Conference, Limehouse Town Hall, London, 10–11 April 2008, at http://www.pnuk.org.uk/bluefence.htm [accessed 29 December 2014]. See also I. Sinclair, 'Growing a New Piece of the City: Designing a Legacy for 21st Century London', Talk given at 21st Century London event, University College London Urban Laboratory, London, 17 November 2009.

8. M. Moorcock, 'Introduction', in I. Sinclair, *Lud Heat and Suicide Bridge* (London: Vintage, 1999), p. 3.
9. S. Pile, 'Introduction', in S. Pile and M. Keith (eds), *Geographies of Resistance* (London: Routledge, 1997), p. 16. See M. de Certeau, *The Practice of Everyday Life* (Berkeley, CA: University of California Press, 1984), pp. xviii, 18, 34–9.
10. S. Pile, 'Introduction', in S. Pile and M. Keith (eds), *Geographies of Resistance* (London: Routledge, 1997), p. 16. See M. de Certeau, *The Practice of Everyday Life* (Berkeley, CA: University of California Press, 1984), p. xiv.
11. G. Poynter and I. MacRury, 'Preface', in G. Poynter and I. MacRury (eds), *Olympic Cities: 2012 and The Remaking of London* (London: Ashgate, 2009), pp. xiii–xvi, on p. xv.
12. The phrase emerges while disclosing emotion on the passing of J. G. Ballard. I. Sinclair, 'Upriver. Review of *Thames: Sacred River* by Peter Ackroyd', *London Review of Books*, 31:12 (25 June 2009), pp. 5–10, on p. 10.
13. J. M. Jacobs, 'A Geography of Big Things', *Cultural Geographies*, 13:1 (2006), pp. 1–27, on p. 3. For the 'goings-on' of the thing realm see J. Baudrillard, *Revenge of the Crystal: Selected Writings on the Modern Art Object and Its Destiny: 1968–1983* (London: Pluto, 1999).
14. M. de Certeau, *The Practice of Everyday Life* (Berkeley, CA: University of California Press, 1984), p. 105.
15. I. Sinclair, 'The Olympics Scam: The Razing of East London', *London Review of Books*, 30:12 (19 June 2008), pp. 17–23, on pp. 18, 20.
16. P. Ackroyd, *First Light* (New York: Grove Press, 1999), p. 99.
17. I. Sinclair, 'Water Walks', in G. Evans and D. Robson (eds), *Towards Re-Enchantment* (London: ArtEvents, 2010), pp. 13–28, on p. 26.
18. D. Harvey, *The Condition of Postmodernity: An Enquiry Into the Origins of Cultural Change* (Oxford: Blackwell, 1989), pp. 147, 240, 260–83, 298, 300.
19. Most notably, I. Sinclair, *Ghost Milk: Recent Adventures Among the Future Ruins of London on the Eve of the London Olympics* (London: Faber and Faber, 2012).
20. I. Sinclair, 'Olympics Scam: The Razing of East London', *London Review of Books*, 30:12 (2008), pp. 17–23, on p. 17.
21. M. Foucault, *The Archaeology of Knowledge*, trans. A. M. Sheridan Smith (London: Tavistock, 1972), p. 130.
22. Sinclair, 'The Olympics Scam', p. 17.
23. Sinclair, 'The Olympics Scam', p. 18.
24. Sinclair has claimed that local residents in Hackney have been threatened with court injunctions for complaining about levels of Thorium in the water table, following Olympic site developments disturbing pollutants from long forgotten watch factories. A journalist at the Hackney Citizen used the Freedom of Information Act (2000) to disclose the fact that the Mayor of Hackney banned Sinclair from 'public buildings' due to his outspoken critique of the development; the BBC have reported that the Olympic Development Agency (ODA) have contracts with all organizations involved in the building of the Olympics that prevents firms from publicly communicating details of

the work for six years while the ODA maintain the right to investigate their property in lieu of any breach of this contract. See BBC, 'Games Firms Sign Gagging Orders', 26 Jan 2009, at http://news.bbc.co.uk/1/hi/england/london/7851358.stm.
25. P. Harries-Jones, 'The Self-Organizing Polity: An Epistemological Analysis of Political Life by Laurent Dobuzinskis', *Canadian Journal of Political Science*, 21:2 (1988), pp. 431–3, on p. 431. See also F. Geyer and J. van der Zouwen (eds), *Sociocybernetics: An Actor-oriented Social Systems Approach*, (Leiden: Nijhoff, 1978); and F. Geyer and J. van der Zouwen (eds), *Sociocybernetics: Complexity, Autopoiesis, and Observation of Social Systems* (Westport, CT: Greenwood Publishing Group, 2001).
26. Harvey, *The Condition of Postmodernity*, p. 88.
27. In addition to the perimeter fence the Olympic security included the following: a fully integrated CCTV system; facial recognition and finger print technology; RAF Reaper pilotless and armed drone aircraft; Royal Navy Daring class Type 45 destroyer; £1bn policing operation with 5000 officers working on event days and 10,000 volunteer police officers. See M. Wells, 'London: The Inclusive Olympiad', *GamesMonitor*, 21 May 2009, at http://www.gamesmonitor.org.uk/node/823.
28. Sinclair, 'The Olympics Scam', p. 19. The ODA used compulsory purchase orders while not consulting with communities: 'residents were given ridiculously short amounts of time to digest and respond to reams, up to 800 pages, of technical reports; some of which could not be downloaded from the Internet and were made "available" to local people at the cost of £500'. L. Porter, 'Exposing the Real Costs of the London Olympics: Lessons for Glasgow 2014', *Scottish Planner* (June 2008), p. 9.
29. S. Elden, *Mapping the Present: Heidegger, Foucault and The Project of a Spatial History* (London: Continuum, 2001), pp. 95–6.
30. I. Sinclair, 'Upriver. Review of *Thames: Sacred River* by Peter Ackroyd', *London Review of Books*, 31:12 (25 June 2009), pp. 5–10, on p. 8.
31. 'For every organism there are limitations and regularities which define what will be learned and under what circumstances this learning will occur. These regularities and patterns become basic premises for the individual adaptation and social organization of any species'; G. Bateson, *Steps to an Ecology of Mind* (Chicago, IL: University of Chicago Press, 1972), p. 422.
32. See Plato, *The Laws*, trans. T. Saunders (London: Penguin, 2004); and N. Weiner, *Cybernetics or Control and Communication in the Animal and the Machine* (Paris: Librairie Hermann and Cie, 1948).
33. Sinclair, 'Upriver', p. 8.
34. Sinclair, 'Upriver', p. 7. The irony is not lost on Sinclair whose 1991 novel, winner of the James Tait Black Award of that year, was titled 'Downriver'.
35. Sinclair, 'Upriver', p. 5.
36. Sinclair, 'Upriver', p. 5.
37. These sites of culture and minds have been identified as the 'three ecologies' by Felix Guattari, which are interactive and interdependent, largely in direct result of a new form of world capital. F. Guattari, *The Three Ecologies* (London: Athlone, 2000).
38. Sinclair, 'Upriver', p. 5.
39. Sinclair, 'Upriver', p. 6.
40. H. Pinter, 'Harold Pinter – Nobel Lecture: Art, Truth, Politics', *Nobelprize.org*, 7 December 2005, at http://nobelprize.org/nobel_prizes/literature/laureates/2005/pinter-lecture.html.
41. Sinclair, 'Upriver', p. 6.

42. Sinclair, 'Upriver'.
43. Hawksmoor, appointed as surveyor in the Office of Works under Sir Christopher Wren shortly after the 1711 Act of Parliament to build fifty churches in London (repeating the project following the Great Fire), was fortunate that the Tories (High Church Party) held office in 1710, a temporal moment which now appears as an interstitial chink in early post-Restoration history that afforded a late bloom of English Baroque before Neo-Palladianism replaced it around 1715.
44. Downes cites the Whig philosopher, the third Earl of Shaftesbury and his attack upon the French taste of Wren's monopoly, while also suggesting that James Leoni's translation of Palladio and the influence of the Scottish architect Colin Campbell contributed to the change in sensibility. K. Downes, *Hawksmoor* (London: Thames and Hudson, 1970), pp. 141–6.
45. I. Sinclair, *Lud Heat: A Book of Dead Hamlets* (London: Albion Village, 1975), p. 29. Ackroyd borrows heavily from Sinclair's inquiry into the relation between sacrifice (the body) and the public sense of lodging (community) for his novel, *Hawksmoor* (1985). For a complete analysis of the relationship between Ackroyd and Sinclair, see A. Murray, *Recalling London: Literature and History in the Work of Peter Ackroyd and Iain Sinclair* (London: Continuum, 2007).
46. I. Sinclair, 'Nicholas Hawksmoor and His Churches', *Flesh Eggs, and Scalp Metal* (London: Paladin, 1989), p. 66.
47. Sinclair, 'Nicholas Hawksmoor', p. 59.
48. Sinclair, 'Nicholas Hawksmoor', pp. 18–19. These knots of energy can be witnessed in the contemporaneous graphic novel by A. Moore, *From Hell: Being A Melodrama in Sixteen Parts* (Canada: Second Top Shelf Printing, 2005), illustrated by Eddie Campbell. See especially chapter four, where Sir William Gull, identified with Jack-the-Ripper, maps out a similar deep energy under the employment of Queen Victoria, as he is tasked with the protection of the British throne (he must murder prostitutes who are aware of the birth of an illegitimate child linked to Victoria's grandson, Prince Albert Victor). Hawksmoor is the subject of Ackroyd's novel: P. Ackroyd, *Hawskmoor* (London: Penguin, 1985).
49. Sinclair, 'Nicholas Hawksmoor', p. 60.
50. Guy Debord's practice-based method studies the effects of the environment on individuals. The study does not need to be consciously organized; a heightened psychological response (or affect) to urban experience takes on new representational possibilities e.g. dream-like 'tactics' or styles of imagining the city over the 'concept city' of rational, urbanist discourse. See M. de Certeau, 'Walking in the City', The Practice of Everyday Life (Berkeley, CA: University of California Press, 1984), pp. 91–110.
51. I. Sinclair, *Lights Out for the Territory* (London: Granta, 1997), p. 147.
52. De Quincey cited in Sinclair, 'Nicholas Hawksmoor', p. 65.
53. D. Defoe, *A Tour Thro' the Whole Island of Great Britain: Divided into Circuits or Journies, Giving a Particular and Diverting Account of Whatever is Curious and Worth Observation*, 3 vols (London: Printed and sold by G. Strahan, 1724–7).
54. Sinclair's premise in the afterword to Gill has Blakean overtones, too: that an artistic response to Hackney's present is 'a wake ... for a city that has *not yet been built*'; I. Sinclair, 'Diving in to Dirt', in S. Gill, *Archaeology in Reverse* (London: n. p., 2007), p. 6.
55. M. Bakhtin, 'Forms of Time and of the Chronotope in the Novel', *The Dialogic Imagination*, trans. C. Emerson and M. Holquist (Austin, TX: University of Texas Press, 1981), pp. 84–258, on p. 250. See also P. Smethurst, *The Postmodern Chronotope: Read-*

ing Space and Time in Contemporary Fiction (Amsterdam: Rodopi, 2005), p. 15.
56. E. Hobsbawn and T. Ranger (eds), *The Invention of Tradition* (Cambridge: Cambridge University Press, 1983); J. Clarke, S. Hall, T. Jefferson and B. Roberts, 'Subcultures, Cultures and Class', in S. Hall and T. Jefferson (eds), *Resistance through Rituals: Youth Subcultures in Post-War Britain* (London: Hutchison, 1976), pp. 9–74, on p. 53; and P. Jackson, *Maps of Meaning: An Introduction to Cultural Geography* (London: Unwin Hyman, 1989).
57. Contemporaneous Palladian interior walls would be 'dead'; Downes, *Hawksmoor*, p. 148.
58. Downes, *Hawksmoor*, pp. 137–8. Downes also argues that St George and St Anne's are 'linked in the architect's mind'; Downes, *Hawksmoor*, p. 117.
59. Sinclair, 'The Olympics Scam', p. 19.
60. Sinclair, *Lights Out*, p. 241.
61. Sinclair, *Lights Out*, p. 243.
62. D. Massey, *For Space* (London: Sage, 2005), p. 148.
63. Sinclair, 'Nicholas Hawksmoor', p. 67.
64. St Anne's masons did not realize the full potential of Hawksmoor's design; it stands as a reminder of that which 'if ever given the chance to come to fruition ... [would become] an overt statement of high ritual, a claim to kinship'; Sinclair, 'Nicholas Hawksmoor', p. 69. Unrealized potential and deep sources combine as Sinclair reads Hawksmoor's paradoxical 'baroque overview' as 'ordered mapping', and that the resource for his churches, Portland stone of Dorset, instances the 'reverse polarity' of 'sacred markers' dispersed through geography and 'hide all the cathedrals that Hawksmoor imagined, the unachieved London of the mind'; Sinclair, *Lights Out*, pp. 187, 232.
65. Sinclair, *Lights Out*, p. 41.
66. Sinclair, *Lights Out*, p. 243.
67. I. Sinclair, 'Upriver. Review of *Thames: Sacred River* by Peter Ackroyd', *London Review of Books*, 31:12 (25 June 2009), pp. 5–10, on p. 5.
68. J. Goss, 'Modernity and Post-Modernity in the Retail Landscape', in K. Anderson and F. Gale, *Inventing Places: Studies in Cultural Geography* (London: Longman, 1992), pp. 159–77, on p. 162.
69. Sinclair, 'Upriver', p. 5.
70. Goss, 'Modernity and Post-Modernity in the Retail Landscape', p. 165.
71. Goss, 'Modernity and Post-Modernity in the Retail Landscape', p. 165. See also C. Jencks, *The Language of Post-Modernist Architecture* (London: Academy Editions, 1987), p. 15.
72. I. Sinclair, *Lights Out for the Territory* (London: Granta, 1997), p. 41.
73. Sinclair, 'Olympics Scam', p. 20.
74. The social is situated between the political and the economic. The political tends to absorb the social, the economic to destroy it. More precisely, the social (otherwise known as civil society) is put between the economic (the base) and the political (the superstructure). H. Lefebvre, *De L'etat*, 4 vols (Paris: UGE, 1976–8), vol. 2, p. 198.
75. E. Grosz, 'Bodies/ Cities', in B. Colomina (ed.), *Sexuality and Space* (Princeton, NJ: Princeton Architectural Press, 1992), pp. 241–54, on pp. 252–3.
76. M. Foucault, 'The Question of Geography', in C. Gordon (ed.), *Power/Knowledge* (New York: Pantheon Books, 1980), p. 70.
77. See P. Morris, *Meaning and Action* (London: Routledge and Kegan Paul, 1987).
78. See H. Ligget and D. C. Perry (eds), *Spatial Practices: An Introduction* (London: Sage,

1995), pp. 1–12, on p. 6.

3 Heterotopia and Placelessness in Brian Chikwava's *Harare North*

1. M. Foucault, 'Of Other Spaces', trans. J. Miskowiec, *Diacritics*, 16:1 (Spring 1986), pp. 22–7, on p. 24.
2. M. Foucault, *The Order of Things* (London: Tavistock Publications, 1970), pp. xv–xxiv.
3. E. Soja, *Thirdspace* (Oxford: Blackwell, 1996), p. 170.
4. H. Heynen, 'Heterotopia Unfolded?', in M. Dehaene and L. De Cauter (eds), *Heterotopia and the City* (Abingdon: Routledge, 2008), pp. 311–23, on p. 320.
5. P. Johnson, 'Unravelling Foucault's "Different Spaces"', *History of the Human Sciences*, 19:4 (2006), pp. 75–90, on p. 84.
6. Foucault, 'Of Other Spaces', p. 24.
7. Foucault, 'Of Other Spaces', p. 25.
8. Foucault, 'Of Other Spaces', p. 24.
9. B. Chikwava, *Harare North* (London: Jonathan Cape, 2009), p. 225.
10. Johnson, 'Unraveling Foucault's "Different Spaces"', p. 86.
11. Johnson, 'Unraveling Foucault's "Different Spaces"', pp. 86–7.
12. Chikwava, *Harare North*, p. 17.
13. Foucault, 'Of Other Spaces', p. 22.
14. Chikwava, *Harare North*, p. 4.
15. Chikwava, *Harare North*, p. 8.
16. Soja, *Thirdspace*, p. 171.
17. Chikwava, *Harare North*, p. 1.
18. Foucault, *The Order of Things*, p. xviii.
19. J. Derrida, in P. Kamuf (ed.), *Between the Blinds – A Derrida Reader* (New York, London: Harvester Wheatsheaf, 1991), p. 76.
20. Foucault, *The Order of Things*, p. xviii.
21. Chikwava, *Harare North*, p. 179.
22. Chikwava, *Harare North*, p. 114.
23. Chikwava, *Harare North*, p. 9.
24. Foucault, 'Of Other Spaces', p. 27.
25. Chikwava, *Harare North*, p. 14.
26. Chikwava, *Harare North*, p. 15.
27. In *The Lonely Londoners* the character Galahad, unemployed and hungry, cooks and eats a pigeon which he catches by feeding the birds in Trafalgar Square.
28. Chikwava, *Harare North*, p. 16.
29. Chikwava, *Harare North*, p. 11.
30. Chikwava, *Harare North*, p. 33.
31. Chikwava, *Harare North*, p. 132.
32. Chikwava, *Harare North*, p. 114.
33. Foucault, 'Of Other Spaces', p. 25.
34. M. Blanchot, *The Space of Literature* (Lincoln, NE: University of Nebraska Press, 1982), p. 25.
35. Chikwava, *Harare North*, p. 226.
36. H. Sohn, 'Heterotopia: Anamnesis of a Medical Term', in M. Dehaene and L. De Cauter (eds), *Heterotopia and the City* (Abingdon: Routledge, 2008), pp. 41–50, on p. 45.
37. Foucault, 'Of Other Spaces', p. 22.

38. Foucault, 'Of Other Spaces', p. 25.
39. Chikwava, *Harare North*, p. 27.
40. Chikwava, *Harare North*, p. 134.
41. S. Selvon, *The Lonely Londoners* (Harlow: Longman, 1956), p. 141.
42. Selvon, *The Lonely Londoners*, p. 142.
43. Selvon, *The Lonely Londoners*, p. 142.
44. Chikwava, *Harare North*, p. 1.
45. Chikwava, *Harare North*, p. 205.
46. Chikwava, *Harare North*, p. 9.
47. Chikwava, *Harare North*, p. 38.
48. M. Boyer, 'The Many Mirrors of Foucault and their Architectural Reflections', in M. Dehaene and L. De Cauter (eds), *Heterotopia and the City* (Abingdon: Routledge, 2008), pp. 53–73, on p. 62.
49. Boyer, 'The Many Mirrors of Foucault', p. 63.
50. Chikwava, *Harare North*, p. 207.
51. Chikwava, *Harare North*, p. 229.
52. Chikwava, *Harare North*, p. 9.
53. Chikwava, *Harare North*, p. 229.
54. Selvon, *The Lonely Londoners*, pp. 101–10.
55. Chikwava, *Harare North*, p. 82.
56. M. Blanchot, 'Translator's Introduction', *The Space of Literature*, pp. 1–18, on p. 10.
57. Chikwava, *Harare North*, p. 29.
58. Foucault, *The Order of Things*, p. 119.
59. Chikwava, *Harare North*, p. 227.
60. Blanchot, 'Translator's Introduction', *The Space of Literature*, p. 10.
61. Foucault, *The Order of Things*, p. 117.

4 'It's a Freedom Thing': Heterotopias and Gypsy Travellers' Spatiality

1. P. Rabinow, *The Foucault Reader* (New York: Pantheon Books, 1984), p. 252.
2. P. Kabachnik, 'To Choose, Fix, or Ignore Culture? The Cultural Politics of Gypsy and Traveler Mobility in England', *Social & Cultural Geography*, 10:4 (2009), pp. 461–79, on p. 464.
3. Kabachnik, 'To Choose, Fix, or Ignore Culture?', p. 467.
4. R. Shields, *Lefebvre, Love and Struggle: Spatial Dialectics* (London: Routledge, 2005), p. 165.
5. Lefebvre cited in Shields, *Lefebvre, Love and Struggle*, p. 168.
6. G. Deleuze and F. Guattari, *New Nietzsche: Contemporary Styles of Interpretation* (Cambridge, MA: Massachusetts Institute of Technology Press, 1985), p. 146.
7. Kabachnik, 'To Choose, Fix, or Ignore Culture?', p. 465.
8. According to Liegeois and Gheorghe, the term 'Gypsy' is 'used to denote ethnic groups formed by the dispersal of commercial, nomadic and other groups from within India from the tenth century, and their mixing with European and other groups during their Diaspora'. 'Traveller' is 'a member of any of the (predominantly) indigenous European ethnic groups ... whose culture is characterized, inter-alia, by self-employment, occupational fluidity, and nomadism. These groups have been influenced, to a greater or

lesser degree, by ethnic groups of (predominantly) Indian origin with a similar cultural base'. J. P. Liegeois and N. Gheorghe, *Roma/Gypsies: A European Minority* (London: Minority Rights Group Report S., 1995), p. 6. In this chapter I refer to the travelling communities in Britain and employ the term 'Gypsy Traveller'. Whilst acknowledging the heterogeneity of such communities, I am aware that labelling can be problematic; however, 'Gypsy Traveller' thus far is the most widely accepted terminology by travelling communities themselves.

9. This entails semi-structured interviews with 15 Gypsy Traveller women (in this chapter I have used both pseudonyms and real names). Interviews were conducted on sites in England and Scotland between 2011–14. The gendered dimension of this data is mainly dictated by the fact that, as a female researcher, gaining access to women was more feasible.

10. D. Garner, 'Fighters and Lovers, All Misunderstood: 'Gypsy Boy', a Memoir by Mikey Walsh', *New York Times*, 16 February 2012, at http://www.nytimes.com/2012/02/17/books/gypsy-boy-a-memoir-by-mikey-walsh.html [accessed August 2014].

11. This methodological choice, which includes both ethnographic data as well as a published autobiographical account, stems from the reflection that one set of material complements the other. Where interviews tend to focus more on the personal perception of space, and on the ways space is individually experienced, the memoir provides actual descriptions of spaces and places, sites and dwellings.

12. G. Deleuze and F. Guattari, *New Nietzsche: Contemporary Styles of Interpretation* (Cambridge, MA: Massachusetts Institute of Technology Press, 1985), p. 146.

13. S. Shubin, '"Where Can a Gypsy Stop?" Rethinking Mobility in Scotland', *Antipode*, 43:2 (2011), pp. 494–524, on p. 498.

14. P. Kabachnik, 'To Choose, Fix, or Ignore Culture? The Cultural Politics of Gypsy and Traveler Mobility in England', *Social & Cultural Geography*, 10:4 (2009), pp. 461–79, on p. 467.

15. J. Okely, *The Traveller-Gypsies* (Cambridge: Cambridge University Press, 1983).

16. D. Kenrick and C. Clark, *Moving On: The Gypsies and Travellers of Britain* (Hatfield: University of Hertfordshire Press, 1999), p. 29.

17. S. Shubin and K. Swanson, '"I'm an Imaginary Figure": Unravelling the Mobility and Marginalization of Scottish Gypsy Travellers', *Geoforum*, 41:6 (2010), pp. 919–29, on p. 921.

18. Shubin and Swanson, '"I'm an Imaginary Figure"', p. 922.

19. R. Mckinney, 'Views from the Margins: Gypsy/Travellers and the Ethnicity Debate in the New Scotland', *Scottish Affairs*, 42 (2003), p. 13–31, on p. 23.

20. D. Mayall, *Gypsy-Travellers in Nineteenth Century Society* (Cambridge: Cambridge University Press, 1988); A. Fraser, *The Gypsies* (Oxford: Wiley Blackwell, 1992); D. Sibley, *Outsiders in Urban Society* (New York: St. Martin's Press, 1981); C. Clark and M. Greenfileds, *Here to Stay: The Gypsies and Travellers of Britain* (Hatfield: University of Hertfordshire Press, 2006); and P. Kabachnik and A. Ryder, 'Nomadism and the 2003 Anti-Social Behaviour Act: Constraining Gypsy and Traveller Mobilities in Britain', *Romani Studies*, 23:1 (2013), p. 83–106.

21. Shubin and Swanson, '"I'...m an Imaginary Figure"', p. 924.

22. M. Palladino, Interview with Maria, 19 July 2014.

23. M. Palladino, Interview with Lol, 19 July 2014

24. Among prominent examples in the West we can recall Italy's recent crackdown on its nomads with the dismantling of Gypsy sites in the peripheries of Rome; 'Rome to

Dismantle Illegal Camps', *BBC News*, 16 February 2009, at http://news.bbc.co.uk/1/hi/world/europe/7893536.stm [accessed August 2014]. Sarkozy's France also came up with a very controversial policy of expulsion which forcibly removed and deported hundreds of Roma Gypsies – effectively EU citizens – back to Romania; 'France Sends Roma Gypsies Back to Romania', *BBC News*, 20 August 2010, at http://www.bbc.co.uk/news/world-europe-11020429 [accessed August 2014].

25. Kabachnik, 'To Choose, Fix, or Ignore Culture?', p. 46.
26. Deleuze and Guattari, *New Nietzsche*, p. 146.
27. Shubin and Swanson, '"I'm an Imaginary Figure"', p. 921.
28. By 'legal' spaces it is intended authorized encampments either privately owned or run by a city council.
29. M. Foucault, 'Of Other Spaces', trans. J. Miskowiec, *Diacritics*, 16:1 (Spring 1986), pp. 22–7, on p. 22.
30. Lefebvre, cited in R. Shields, *Lefebvre, Love and Struggle: Spatial Dialectics* (London: Routledge, 2005), p. 180.
31. I. Tyler, *Revolting Subjects* (London: Zed Books, 2013), p. 2.
32. G. Deleuze and F. Guattari, *Thousand Plateaus: Capitalism and Schizophrenia* (Minneapolis, MN, and London: University of Minnesota Press, 2005), p. 380.
33. Deleuze and Guattari, *Thousand Plateaus*, p. 380; emphasis original.
34. Deleuze and Guattari, *Thousand Plateaus*, p. 382.
35. M. Foucault, 'Of Other Spaces', trans. J. Miskowiec, *Diacritics*, 16:1 (Spring 1986), pp. 22–7, on p. 25.
36. M. P. Levinson and A. C. Sparkes, 'Gypsy Identity and Orientations to Space', *Journal of Contemporary Ethnography*, 33:6 (2004), pp. 704–34, on p. 713.
37. A. Bancroft, '"No Interest in Land": Legal and Spatial Enclosure of Gypsy-Travellers in Britain', *Space and Polity*, 4:1 (2000), pp. 41–56, on p. 41.
38. Levinson and Sparkes, 'Gypsy Identity and Orientations to Space', p. 723.
39. M. Foucault, 'Theatrum philosophicum', in D. F. Bouchard (ed.), *Language, Counter-memory, Practice* (Ithaca, NY: Cornell University Press, 1977), p. 355.
40. G. Doron, '"Those Marvellous Empty Zones on the Edge of Our Cities": Heterotopia and the "Dead Zone"', in M. Dehaene and L. De Cauter (eds), *Heterotopia and the City: Public Space in a Postcivil Society* (London: Routledge, 2008), pp. 204–7, on p. 204.
41. Doron explains this concept as follows: 'I have been using the term "dead zone" (translated from Hebrew), which was taken from planners' jargon and more specifically from a discussion about a regeneration plan that will be mentioned below, to indicate a gap, if not a total break, between the signifier and the signified; Doron, '"Those Marvellous Empty Zones on the Edge of Our Cities"', p. 204).
42. M. Walsh, *Gypsy Boy* (London: Hodder & Stoughton, 2009), p. 157–86.
43. Deleuze and Guattari, *Thousand Plateaus*, p. 381.
44. Walsh, *Gypsy Boy*, p. 79.
45. Deleuze and Guattari, *Thousand Plateaus*, p. 381.
46. M. Cenzatti, 'Heterotopias of Difference', in M. Dehaene and L. De Cauter (eds), *Heterotopia and the City: Public Space in a Postcivil Society* (London: Routledge, 2008), pp. 75–86, on p. 77.
47. Walsh, *Gypsy Boy*, p. 113.
48. Foucault, 'Of Other Spaces', p. 25.
49. H. Sohn, 'Heterotopia: Anamnesis of a Medical Term', in M. Dehaene and L. De Cauter (eds), *Heterotopia and the City: Public Space in a Postcivil Society* (London:

Routledge, 2008), pp. 41–50, on p. 47.
50. M. Foucault, 'Of Other Spaces', trans. J. Miskowiec, *Diacritics*, 16:1 (Spring 1986), pp. 22–7, on p. 26.
51. J. Okely, *The Traveller-Gypsies* (Cambridge: Cambridge University Press, 1983).
52. C. Karner, 'Theorising Power and Resistance among "Travellers"', *Social Semiotics*, 14:3 (2004), pp. 249–71, on p. 262.
53. M. Greenfields and D. M. Smith, 'Housed Gypsy Travellers, Social Segregation And The Reconstruction Of Communities', *Housing Studies*, 25:3 (2010), pp. 397–412, on p. 406.
54. For Gypsy Travellers 'Gorgia' or 'Gorgio' designates non travelling people.
55. M. P. Levinson and A. C. Sparkes, 'Gypsy Identity and Orientations to Space', *Journal of Contemporary Ethnography*, 33:6 (2004), pp. 704–34, on p. 730.
56. E. Soja, *Thirdspace: Journeys to Los Angeles and Other Real-and-Imagined Places* (Oxford: Blackwell, 1996), p. 67.
57. Channel 4 broadcasted the documentary series *Big Fat Gypsy Weddings* from 2011 till 2014; a controversial programme – for the ways it represents the Gypsy Traveller communities – it sparked debates across the country and ignited a remarkable interest in the subject. BBC TV and Radio also produced several programmes about Gypsy Travellers.
58. D. Papadopoulos, N. Stephenson and V. Tsianos, *Escape Routes: Control and Subversion in the 21st Century* (London: Pluto Press, 2008).
59. I. Tyler, *Revolting Subjects* (London: Zed Books, 2013), p. 11.
60. M. Palladino, Interview with Lol, 19 July 2014.
61. Tyler, *Revolting Subjects*, p. 2.
62. Imogen Tyler identifies among the 'revolting subjects' in contemporary neo-liberal Britain Gypsy Travellers, migrants and refugees, the working classes. Tyler, *Revolting Subjects*.
63. Tyler, *Revolting Subjects*, p. 12.
64. M. Cenzatti, 'Heterotopias of Difference', in M. Dehaene and L. De Cauter (eds), *Heterotopia and the City: Public Space in a Postcivil Society* (London: Routledge, 2008), pp. 75–86, on p. 82.
65. S. Kendall, 'Sites of Resistance: Places on the Margin – The Traveller "Homeplace"', in T. Acton (ed.), *Gypsy Politics and Traveller Identity* (Hatfield: University of Hertfordshire Press, 1997), pp. 70–89, on p. 75.
66. C. Karner, 'Theorising Power and Resistance among "Travellers"', *Social Semiotics*, 14:3 (2004), pp. 249–71, on p. 265.
67. S. Shubin, '"Where Can a Gypsy Stop?" Rethinking Mobility in Scotland', *Antipode*, 43:2 (2011), p. 494–524), p. 498.
68. D. Harvey, *Rebel City* (London: Verso, 2012), p. xxvii.
69. Foucault, 'Of Other Spaces', p. 26.
70. Karner, 'Theorising Power and Resistance among "Travellers"', p. 263.
71. M. Walsh, *Gypsy Boy* (London: Hodder & Stoughton, 2009), p. 77.
72. Walsh, *Gypsy Boy*, pp. 81, 156.
73. M. Foucault, 'Of Other Spaces', trans. J. Miskowiec, *Diacritics*, 16:1 (Spring 1986), pp. 22–7, on p. 23.
74. J. Okely, *The Traveller-Gypsies* (Cambridge: Cambridge University Press, 1983), p. 49.
75. M. P. Levinson and A. C. Sparkes, 'Gypsy Identity and Orientations to Space', *Journal of Contemporary Ethnography*, 33:6 (2004), pp. 704–34, on p. 725.
76. Levinson and Sparkes, 'Gypsy Identity and Orientations to Space', p. 729.

77. M. Palladino, Interview with Maria, 19 July 2014.
78. Foucault, 'Of Other Spaces', p. 27.
79. Foucault, 'Of Other Spaces', p. 27.
80. M. Walsh, *Gypsy Boy* (London: Hodder & Stoughton, 2009), p. 149.
81. Walsh, *Gypsy Boy*, p. 28.
82. Levinson and Sparkes, 'Gypsy Identity and Orientations to Space', p. 719.
83. M. Palladino, Interview with Lol, 19 July 2014.
84. Palladino, Interview with Maria.
85. M. Palladino, Interview with Margaret, 19 July 2014.
86. Levinson and Sparkes, 'Gypsy Identity and Orientations to Space', p. 721.
87. Foucault, 'Of Other Spaces', p. 23–4.
88. Foucault, 'Of Other Spaces', p. 27.
89. P. Johnson, 'Unravelling Foucault's "Different Spaces"', *History of the Human Sciences*, 19:4 (2006), pp. 75–90, on p. 80.
90. G. Deleuze and F. Guattari, *New Nietzsche: Contemporary Styles of Interpretation* (Cambridge, MA: Massachusetts Institute of Technology Press, 1985), p. 146.

5 Heterotopias of Illness

1. L. Tanner, *Lost Bodies: Inhabiting the Borders of Life and Death* (Ithaca, NY: Cornell University Press, 2006), p. 64.
2. H. Sohn, 'Heterotopia: Anamnesis of a Medical Term', in M. Dehaene and L. De Cauter (eds), *Heterotopia and the City: Public Space in a Postcivil Society* (Oxon: Routledge, 2008), pp. 41–50, on p. 48.
3. K. Hetherington, *The Badlands of Modernity: Heterotopia and Social Ordering* (London: Routledge, 1997), p. 7.
4. Some of these texts (*Madonna Swan* is a clear example) raise further questions of power imbalance, ethics and appropriation, which I do not have the space to address here.
5. See K. R. Myers, *The Patient: Global Interdisciplinary Perspectives* (Oxford: Inter-Disciplinary Press, 2009); C. Barker and S. Murray, 'Disabling Postcolonialism: Global Disability Cultures and Democatic Criticsm', *Journal of Literary & Cultural Disability Studies*, 4:3 (2010), pp. 219–36; C. Hooker and E. Noonan, 'Medical Humanities as Expressive of Western Culture', *Medical Humanities*, 37:2 (2011), pp. 79–84; and the conference 'Global Medical Humanities' organized by the Association for Medical Humanities, 2013.
6. V. Woolf, *On Being Ill* (Ashfield: Paris Press, 2002), p. 3.
7. S. Sontag, *Illness as Metaphor and AIDS and Its Metaphors* (London: Penguin, 1991), p. 3; emphasis added.
8. A. Frank, *At the Will of the Body: Reflections on Illness* (New York: Mariner, 2002), p. 129.
9. A. Frank, *The Wounded Storyteller: Body, Illness, and Ethics* (London: University of Chicago Press, 1997), p. 9.
10. In their translation, Dehaene and De Cauter opt for this term (rather than temporal) to stress the 'cyclical' nature of the heterotopia of festivity. This seems appropriate for patients in remission and those who experience relapses, as in the case of mental illness. See M. Foucault, 'Of Other Spaces', trans. M. Dehaene and L. De Cauter, in M. Dehaene and L. De Cauter (eds), *Heterotopia and the City: Public Space in a Postcivil Society* (London: Routledge, 2008), pp. 13–29, on p. 26, n. 24.

11. L. Tanner, *Lost Bodies: Inhabiting the Borders of Life and Death* (Ithaca, NY: Cornell University Press, 2006), p. 64.
12. H. Sohn, 'Heterotopia: Anamnesis of a Medical Term', in M. Dehaene and L. De Cauter (eds), *Heterotopia and the City: Public Space in a Postcivil Society* (Oxon: Routledge, 2008), pp. 41–50, on p. 41.
13. M. Foucault, 'Of Other Spaces', trans. J. Miskowiec, *Diacritics*, 16:1 (Spring 1986), pp. 22–7, on p. 23.
14. P. Johnson, 'Unravelling Foucault's "Different Spaces"', *History of the Human Sciences*, 19:4 (2006), pp. 75–90, on p. 84.
15. M. St Pierre, *Madonna Swan: A Lakota Woman's Story* (Norman, OK: University of Oklahoma Press, 1991), p. 65.
16. L. De Cauter and M. Dehaene, 'The Space of Play: Towards a General Theory of Heterotopia', in M. Dehaene and L. De Cauter (eds), *Heterotopia and the City: Public Space in a Postcivil Society* (Oxon: Routledge, 2008), pp. 87–102, on p. 92; emphasis added.
17. L. Diedrich, *Treatments: Language, Politics, and the Culture of Illness* (Minneapolis, MN: University of Minnesota Press, 2007), p. 19; emphasis added.
18. St Pierre, *Madonna Swan*, p. 40.
19. St Pierre, *Madonna Swan*, p. 191, n. 73.
20. St Pierre, *Madonna Swan*, p. 63.
21. S. Sontag, *Illness as Metaphor and AIDS and Its Metaphors* (London: Penguin, 1991), p. 18.
22. M. Foucault, 'Of Other Spaces', trans. J. Miskowiec, *Diacritics*, 16:1 (Spring 1986), pp. 22–7, p. 26.
23. St Pierre, *Madonna Swan*, p. 80.
24. St Pierre, *Madonna Swan*, p. 74.
25. St Pierre, *Madonna Swan*, p. 76.
26. St Pierre, *Madonna Swan*, p. 79.
27. See L. De Cauter and M. Dehaene, 'Heterotopia in a Postcivil Society', in M. Dehaene and L. De Cauter (eds), *Heterotopia and the City: Public Space in a Postcivil Society* (Oxon: Routledge, 2008), pp. 3–9, on p. 5.
28. Foucault, 'Of Other Spaces', p. 26.
29. St Pierre, *Madonna Swan*, p. 71.
30. St Pierre, *Madonna Swan*, p. 74.
31. Foucault, 'Of Other Spaces', p. 26.
32. Diedrich, *Treatments,* p.18.
33. St Pierre, *Madonna Swan*, p. 106.
34. St Pierre, *Madonna Swan*, p. 108.
35. Though I am using it in a different context, I have borrowed the title of this section from a phrase in L. De Cauter and M. Dehaene, 'The Space of Play: Towards a General Theory of Heterotopia', in M. Dehaene and L. De Cauter (eds), *Heterotopia and the City: Public Space in a Postcivil Society* (Oxon: Routledge, 2008), pp. 87–102, on p. 92.
36. A. Lorde, 'A Burst of Light: Living with Cancer', *The Audre Lorde Compendium: Essays, Speeches and Journals* (London: Pandora, 1996), pp. 269–335, on p. 297.
37. M. D. Jones, Interview with Elizabeth Alexander, at http://www.elizabethalexander.net/Meta%20Jones%20Interview.pdf [accessed 10 April 2013].
38. Lorde, 'A Burst of Light', p. 292.
39. Lorde, 'A Burst of Light', p. 295.
40. Lorde, 'A Burst of Light', p. 294.

41. Lorde, 'A Burst of Light', p. 296.
42. Lorde, 'A Burst of Light', p. 296.
43. Lorde, 'A Burst of Light', p. 295.
44. Lorde, 'A Burst of Light', p. 297.
45. H. Heynen, 'Heterotopia Unfolded?', in M. Dehaene and L. De Cauter (eds), *Heterotopia and the City: Public Space in a Postcivil Society* (Oxon: Routledge, 2008), pp. 311–23, on p. 315.
46. Lorde, 'A Burst of Light', p. 303.
47. Lorde, 'A Burst of Light', p. 295.
48. A. Lorde, *Zami: A New Spelling of My Name* (London: Sheba, 1982), p. 71.
49. Lorde, 'A Burst of Light', p. 299.
50. Lorde, 'A Burst of Light', p. 301.
51. B. Ochsner, 'Barbara's Blog', *Lukas Klinik*, at http://www.klinik-arlesheim.ch/de/nc/medizinische-angebote/onkologie/erfahrungsberichte/barbaras-blog/?sword_list[]=blog [accessed 10 April 2013].
52. S. Low, 'The Gated Community as Heterotopia', in M. Dehaene and L. De Cauter (eds), *Heterotopia and the City: Public Space in a Postcivil Society* (Oxon: Routledge, 2008), pp.153–64, on p. 162.
53. A. Lorde, 'The Cancer Journals', *The Audre Lorde Compendium: Essays, Speeches and Journals* (London: Pandora, 1996), pp. 3–63, on p. 33.
54. M. Foucault, 'Of Other Spaces', trans. J. Miskowiec, *Diacritics*, 16:1 (Spring 1986), pp. 22–7, on p. 25.
55. L. Burke, 'The Poetry of Dementia: Art, Ethics and Alzheimer's Disease in Tony Harrison's *Black Daisies for the Bride*', *Journal of Literary Disability*, 1:1 (2007), pp. 61–73, on p. 63.
56. L. Grant, *Remind Me Who I Am, Again* (London: Granta, 1998), p. 184.
57. Grant, *Remind Me Who I Am, Again*, p. 174.
58. Grant, *Remind Me Who I Am, Again*, p. 175.
59. L. De Cauter and M. Dehaene, 'The Space of Play: Towards a General Theory of Heterotopia', in M. Dehaene and L. De Cauter (eds), *Heterotopia and the City: Public Space in a Postcivil Society* (Oxon: Routledge, 2008), pp. 87–102, on p. 93.
60. Grant, *Remind Me Who I Am, Again*, p. 224.
61. Grant, *Remind Me Who I Am, Again*, p. 175.
62. Foucault, 'Of Other Spaces', p. 25.
63. Grant, *Remind Me Who I Am, Again*, p. 203.
64. Grant, *Remind Me Who I Am, Again*, p. 204.
65. Grant, *Remind Me Who I Am, Again*, p. 262.
66. Grant, *Remind Me Who I Am, Again*, p. 219.
67. V. Woolf, *On Being Ill* (Ashfield: Paris Press, 2002), p. 12.
68. Grant, *Remind Me Who I Am, Again*, p. 196.
69. Grant, *Remind Me Who I Am, Again*, p. 200.
70. Grant, *Remind Me Who I Am, Again*, p. 205.
71. J. D. Faubion, 'Heterotopia: An Ecology', in M. Dehaene and L. De Cauter (eds), *Heterotopia and the City: Public Space in a Postcivil Society* (Oxon: Routledge, 2008), pp. 31–9, on p. 32.
72. Grant, *Remind Me Who I Am, Again*, p. 216.
73. Grant, *Remind Me Who I Am, Again*, p. 204.
74. M. Cenzatti, 'Heterotopias of Difference', in M. Dehaene and L. De Cauter (eds),

Heterotopia and the City: Public Space in a Postcivil Society (Oxon: Routledge, 2008), pp. 75–85, on p. 77.
75. Foucault, 'Of Other Spaces', pp. 25–6.
76. Grant, *Remind Me Who I Am, Again*, p. 223.
77. D. Muzzio and J. Muzzio-Rentas, '"A Kind of instinct": The Cinematic Mall as Heterotopia', in M. Dehaene and L. De Cauter (eds), *Heterotopia and the City: Public Space in a Postcivil Society* (Oxon: Routledge, 2008), pp. 137–150, on p. 140.
78. Muzzio and Muzzio-Rentas, '"A Kind of instinct"', p. 144.
79. Muzzio and Muzzio-Rentas, '"A Kind of instinct"', p. 146.
80. Grant, *Remind Me Who I Am, Again*, p. 287.
81. M. Foucault, *The Order of Things: An Archaeology of the Human Sciences* (New York, NY: Vintage, 1994), p. xviii; emphasis original.
82. See, for example, S. Egan, *Mirror Talk: Genres of Crisis in Contemporary Autobiography* (London: University of North Carolina Press, 1999); and S. M Squier, *Liminal Lives: Imagining the Human at the Frontiers of Biomedicine* (Durham, NC: Duke University Press, 2004).
83. A. Radley, *Works of Illness: Narrative, Picturing, and the Social Response to Serious Disease* (Ashby-de-la-Zouch: InkerMen Press, 2009), p. 38.
84. K. Hetherington, *The Badlands of Modernity: Heterotopia and Social Ordering* (London: Routledge, 1997), p. 46.

6 Writing the Littoral

1. J. Conrad, 'A Smile of Fortune', *Twixt Land and Sea* (1911; London: Penguin, 1990), pp. 7–78, on p. 77.
2. M. Foucault, 'Of Other Spaces', trans. J. Miskowiec, *Diacritics*, 16:1 (Spring 1986), pp. 22–7, on p. 25.
3. His first novel *Almayer's Folly* was published in 1895; J. Conrad, *Almayer's Folly* (London: T. Fisher Unwin, 1895).
4. J. Conrad, 'A Smile of Fortune', *'Twixt Land and Sea* (1911; London: Penguin, 1990), pp. 7–78, on p. 9.
5. Conrad, 'A Smile of Fortune', p. 35.
6. Conrad, 'A Smile of Fortune', p. 64.
7. There are other hints. The elder Jacobus is described as 'swarthy' (Conrad, 'A Smile of Fortune', p. 28), so perhaps there is an insinuation of racial uncertainty in planter cultures on the margin of European empires.
8. Conrad, 'A Smile of Fortune', p. 34.
9. Z. Najder, *Joseph Conrad: A Chronicle* (Cambridge: Cambridge University Press, 1983), p. 107. In his Author's Note on the story's republication in volume form, Conrad issues this disavowal: 'Notwithstanding their autobiographical form the above two stories ['A Smile of Fortune' and 'The Secret Sharer'] are not the record of personal experience'; J. Conrad, 'Author's Note', *'Twixt Land and Sea* (included in the 1990 Penguin edition, dated 1920), pp. 3–5, on p. 4. Najder, though, shows that there is some overlap between biography and fiction.
10. Conrad, 'Author's Note', *'Twixt Land and Sea*, 1920, p. 3.
11. Conrad, *'Twixt Land and Sea*, p. 53.
12. Conrad, *'Twixt Land and Sea*, p. 27.
13. Conrad, *'Twixt Land and Sea*, p. 34.

14. Conrad, *'Twixt Land and Sea*, p. 29.
15. Conrad, *'Twixt Land and Sea*, p. 34.
16. Conrad, *'Twixt Land and Sea*, p. 56.
17. These are Conrad, *'Twixt Land and Sea*, pp. 13, 21, 32, 37, 43, 48, 72.
18. Conrad, *'Twixt Land and Sea*, p. 11.
19. We know from Nadjer that Conrad may have been involved in a surreptitious affair with a seventeen year-old girl called Alice Shaw – whose father owned the only rose garden in Port Louis. Najder, *Joseph Conrad: A Chronicle*, p. 107.
20. K. Blixen, *Out of Africa* (London: Penguin, 2001), p. 132.
21. To name only the most famous: R. F. Burton, *The Lake Regions of Central Equatorial Africa, With Notices of the Lunar Mountains and the Sources of the White Nile; Being the Results of an Expedition Undertaken Under the Patronage of Her Majesty's Government and the Royal Geographical Society of London, in the Years 1857–1859* (London: John Murray, 1859); J. Hanning Speke, *Journal of the Discovery of the Source of the Nile* (Edinburgh: William Blackwood and Sons, 1863); and D. Livingstone, *The Last journals of David Livingstone in Central Africa, From 1865 to his Death: Continued by a Narrative of his Last Moments and Sufferings, Obtained from his Faithful Servants, Chuma and Susi*, ed. H. Waller, 2 vols (London: J. Murray, 1874).
22. He wrote over forty books and hundreds of articles on exploration, travel, poetry and military history. He also translated *The Book of A Thousand and One Nights* (1885) in 10 volumes, as well as Camoes's *Lusiads* (1880).
23. 'Swaheli' is the Somali form of Waswahili, and it is interesting that Blixen uses this form, presumably acquired from her Somali informant on the subject, her 'butler' Farah Aden, whose own racism perhaps comes through in Blixen's malign description of the Waswahili as 'slave-hearted'; Blixen, *Out of Africa*, p. 132.
24. Blixen, *Out of Africa*, pp. 111–12.
25. The trans-Atlantic crossing from the Cape Verde Islands to within sight of Brazil and then back across to make landfall on an unknown African coast took Vasco Da Gama's fleet three months out of sight of land.
26. See J. Middleton, *The World of the Waswahili* (New Haven, CT: Yale University Press, 1992); M. Horton, *Shanga* (Cambridge: Cambridge University Press, 1980); and D. Nurse and T. Spear, *The Swahili: Reconstructing the History and Language of an African Society, 800–1500* (Philadelphia, PA: University of Pennsylvania Press, 1985).
27. Blixen, *Out of Africa*, p. 112.
28. V. S. Naipaul had travelled in the Congo in 1975 and wrote about it. He published V. S. Naipaul, 'A New King for the Congo', *New York Review of Books*, 26 June 1975, which was later collected in V. S. Naipaul, *The Return of Eva Peron, with the Killings in Trinidad* (London: Penguin, 1980). That volume included 'Conrad's Darkness' also first published in *New York Review of Books*: V. S. Naipaul, 'Conrad's Darkness', *New York Review of Books*, 17 October 1974.
29. V. S. Naipaul, 'Conrad's Darkness', *The Return of Eva Peron* (London: Penguin, 1980), pp. 197–218, on p. 206.
30. Naipaul, 'Conrad's Darkness', 1980, p. 209.
31. J. Conrad, 'Karain', *Tales of Unrest* (London: Fisher Unwin, 1898). 'It appeared to us as a land without memories, regrets and hopes; a land where nothing could survive the coming of the night, and where each sunrise, like a dazzling act of special creation, was disconnected from the eve and the morrow'; J. Conrad, 'Karain', *Tales of Unrest* (Harmondsworth: Penguin, 1985), p. 14.

32. V. S. Naipaul, *A Bend in the River* (London: Vintage, 1989), p. 15.
33. Naipaul, *A Bend in the River*, p. 15.
34. According to Naipaul's biographer, P. French, *The World Is What It Is* (London: Picador, 2008), the model for the Domain in *A Bend in the River* is the Makerere University campus in Kampala, where Naipaul was a Visiting Fellow in the early 1960s.
35. Naipaul, *A Bend in the River*, p. 11.
36. In other words, they were Indians first, a thesis Naipaul would develop more fully in V. S. Naipaul, *Among the Believers: An Islamic Journey* (London: Andre Deutsch, 1981), with his argument that the 'converted people' of Asia, in Iran, Pakistan, Indonesia are only superficially Muslim because they never quite gave up their pre-Islamic religions and cultures, even though Arab 'imperial Islam' required them to do so.
37. Naipaul, *A Bend in the River*, p. 11.
38. Naipaul, *A Bend in the River*, p. 11.
39. See G. Casale, *The Ottoman Age of Exploration* (Oxford: Oxford University Press, 2010) for an extensive survey of non-European Indian Ocean exploration texts.
40. See Naipaul's discussion of Ghandi's sense of mythic time in V. S Naipaul, 'Indian Autobiographies', *New Statesman*, 29 January 1965, in V. S. Naipaul, *The Overcrowded Baracoon, and Other Articles* (London: Andre Deutsch, 1972), pp. 55–60. See also the figure of Gurudeva in V. S. Naipaul, *The Mimic Men* (London: Andre Deutsch, 1967).
41. See the opening section of P. Theroux, *Sir Vidia's Shadow* (London: Hamish Hamilton, 1998).
42. Naipaul, *A Bend in the River*, p. 17.
43. Naipaul, *A Bend in the River*, p. 18.
44. French, *The World Is What It Is*, p. 259. It is an absurd moment in V. S. Naipaul's career, when he took a job between the publication of V. S. Naipaul, *The Mystic Masseur* (Harmondsworth: Penguin, 1957) and V. S. Naipaul, *Miguel Street* (London: Andre Deutsch, 1957). He worked for a magazine called *Concrete Quarterly*, a publication of the building trade. He met Sondhi on a course in Slough to learn about 'construction and architecture'.
45. French, *The World Is What It Is*, p. 261.
46. See J. de Vere Allen, *Swahili Origins* (Athens, OH: Ohio University Press, 1993).
47. Naipaul, *A Bend in the River*, p. 13.
48. O. Mannoni, *Prospero and Caliban: The Psychology of Colonisation*, 2nd revised edn (Ann Arbor, MI: University of Michigan Press, 1990).
49. Naipaul, *A Bend in the River*, pp. 13–14.
50. Naipaul, *A Bend in the River*, p. 21.
51. Naipaul, *A Bend in the River*, p. 37.
52. Naipaul, *A Bend in the River*, p. 45.
53. Naipaul, *A Bend in the River*, p. 61.
54. Naipaul, *A Bend in the River*, p. 42.
55. S. Naipaul, *North of South* (London: Penguin, 1980), p. 13.
56. At the time of independence, 'Asians' in 'British East Africa' were offered the choice of a British passport. A substantial number accepted the offer. Post-independence East African countries put pressure on Asian mercantile interests, short of expulsion. With typically outrageous brutality, President Amin expelled Ugandan Asians in August 1972, giving them 90 days to dispose of their property and leave.
57. Naipaul, *North of South*, p. 100.
58. Naipaul, *North of South*, p. 104.

59. Naipaul, *North of South*, p. 73.
60. Naipaul, *North of South*, p. 107.
61. V. S. Naipaul, *A Bend in the River* (London: Vintage, 1989), p. 18.
62. Naipaul, *North of South*, p. 120.
63. Naipaul, *North of South*, p. 19.
64. Naipaul, *North of South*, p. 22.
65. M. Mwangi, *Going Down River Road* (London: Heinemann Education, 1976).
66. Naipaul, *North of South*, p. 41.
67. Naipaul, *North of South*, p. 43.
68. Naipaul, *North of South*, p. 54.
69. Naipaul, *North of South*, p. 47.
70. Naipaul, *North of South*, p. 272.
71. Naipaul, *North of South*, p. 177.
72. Naipaul, *North of South*, p. 178.
73. Naipaul, *North of South*, p. 188.
74. Naipaul, *North of South*, p. 186.
75. Naipaul, *North of South*, p. 188.
76. Naipaul, *North of South*, p. 193.

7 Heterotopia and the Critical Cut

1. P. Chatterjee, *The Politics of the Governed: Reflections on Popular Politics in Most of the World* (New York: Columbia University Press, 2006), p. 6–7.
2. M. Foucault, 'Of Other Spaces', trans. J. Miskowiec, *Diacritics*, 16:1 (Spring 1986), pp. 22–7, on p. 26.
3. M. Foucault, *The Order of Things: Archaeology of the Human Sciences* (London: Routledge, 2001), pp. xxiv–xxvi.
4. H. Lefebvre, *The Production of Space* (Oxford: Wiley-Blackwell, 1991).
5. C. Conelli, 'Per una storia postcoloniale del Mezzogiorno d'Italia' (MA dissertation, University of Naples, 2012).
6. I. Chambers, 'The "Unseen Order": Religion, Secularism and Hegemony', in N. Srivastava and B. Bhattacharya (eds), *The Postcolonial Gramsci* (New York: Routledge, 2012).
7. S. Mezzadra and B. Neilson, *Border as Method, or, the Multiplication of Labor* (Durham, NC: Duke University Press, 2013).
8. D. Chakrabarty, *Provincializing Europe: Postcolonial Thought and Historical Difference* (Princeton, NJ: Princeton University Press, 2001), p. 27.
9. D. Walcott, 'The Sea is History', in D. Walcott, *Collected Poems* (London: Faber and Faber, 1992), pp. 364–7, on p. 364.
10. A. Sekula, *Fish Story* (Düsseldorf: Richter Verlag, 2003).
11. C. Schmitt, *Land and Sea* (Washington D.C.: Plutarch Press, 1997).
12. R. D. G. Kelley, *Race Rebels: Culture, Politics, and the Black Working Class* (New York, NY: The Free Press, 1994), p. 1.
13. I. Chambers, 'Adrift and Exposed: The Art of Isaac Julien', *Location, Borders and Beyond* (Charleston, SC: CreateSpace, 2012).
14. R. Guha, *History at the Limit of World-History* (New York: Columbia University Press, 2002), p. 12.
15. Guha, *History at the Limit of World-History*, p. 24.
16. Guha, *History at the Limit of World-History*, p. 8.

17. B. Ettinger, *The Matrixial Borderspace* (Minneapolis, MN: University of Minnesota Press, 2006).
18. S. Goodman, *Sonic Warfare: Sound, Affect and the Ecology of Fear* (Cambridge, MA: Massachusetts Institute of Technology Press, 2010), p. 82.
19. G. Deleuze and F. Guattari, *A Thousand Plateaus: Capitalism and Schizophrenia* (London: Continuum, 2004).
20. A. Gramsci, *Selections from the Prison Notebooks of Antonio Gramsci*, ed. Q. Hoare and G. Nowell-Smith (New York: International Publishers, 1971), p. 324.
21. J. Derrida, *Archive Fever: A Freudian Impression* (Chicago, IL: University of Chicago Press, 1998), p. 36.
22. K. Eshun, 'Drawing the Forms of Things Unknown', in K. Eshun and A. Sagar (eds), *The Ghosts of Songs. The Film Art of the Black Audio Collective* (Liverpool: University of Liverpool Press, 2007), p. 78.
23. E. Said, *Musical Elaborations* (London: Vintage, 1992).
24. B. Nettl, 'Interpreting the Musical Map: Recalling Some Neglected Classics', in P. V. Bohlman and M. Sorce Keller (eds), *Antropolgia della musica nelle culture mediterranee. Interpretazione, performance, identità/Muscial Anthology in Mediterranean Cultures: Interpretation, Performance, Identity* (Bologna: CLUEB, 2009).
25. G. Léothaud and B. Lortat-Jacob, 'La voix méditerranéenne. Une identité problématique', in L. Charles-Dominique and J. Cler (eds), *La Vocalité dans le pays d'Europe méridionale et dans les basin méditerranéen* (Saint-Jouin-de-Milly: Modal Éditions, 2000).
26. L. K. Johnson, *Dread, Beat and Blood* (London: Bogle-l'Ouverture Press, 1975).
27. J. Henriques, *Sonic Bodies: Reggae Sound Systems, Performance Techniques, and Ways of Knowing* (London: Continuum, 2011), p. 122.
28. A. J. Racy, *Making Music in the Arab World: The Culture and Artistry of Tarab* (Cambridge: Cambridge University Press, 2004), p. 9.
29. R. Connell, *Southern Theory* (Cambridge: Polity, 2007).
30. G. Didi-Huberman, *Devant el temps. Histoire de l'art et anachronisme des images* (Paris: Éditions du Minuit, 2000).
31. R. Guha, *History at the Limit of World-History* (New York: Columbia University Press, 2002), p. 25.
32. G. Bhambra, *Rethinking Modernity: Postcolonialism and the Sociological Imagination* (London: Palgrave Macmillan, 2009).
33. Bhambra, *Rethinking Modernity*, p. 81.

8 'L'Asile Flottant': Modernist Reflections by the Armée du Salut and le Corbusier on the Refuge/Refuse of Modernity

1. M. Foucault, 'Of Other Spaces', trans. J. Miskowiec, *Diacritics*, 16:1 (Spring 1986), pp. 22–7. In the earlier audio version of the text, M. Foucault, *Le corps utopique, Les hétérotopies* (1966; Paris: Nouvelles éditions lignes, 2009), p. 36, Foucault had said: 'a boat is ... *fatally* exposed ['*livréfatalement*', handed over to] the sea'. The differential quality of finite lives is perhaps the main topic opened up by this exploration of the neglected 'heterotopia'; that was the '*asile flottant*'. See Z. Bauman, *Wasted Lives: Modernity and Its Outcasts* (Cambridge: Polity Press, 2004) on 'wasted lives'.
2. R. Schérer, *Utopies nomades* (Dijon: Presses du réel, 2009), p. 17. All translations from

the French are my own, unless otherwise stated.
3. It needs to be noted that the barge formed the first stage of the programme for the City of Refuge as a whole. Accordingly, Albin Peyron wrote in *En avant*, 1 June 1929, Bibliothèque de la Société de l'Histoire du Protestantisme Français, Paris, that the 'floating asylum': 'will be in constant contact with the city of refuge and it must be organised straightaway'. Given its pivotal importance in the overall City of Refuge programme, it is remarkable that this project has been so little discussed. However, that being said, maybe we should not be surprised: as we know from their biological origins, heterotopias, which for society as a whole signal a troubling zone of 'crisis' or a 'deviation', can be reabsorbed by the overpowering body politic, thereby condemning them to oblivion; M. Foucault, *Le corps utopique, Les hétérotopies* (1966; Paris: Nouvelles éditions lignes, 2009), p. 27.
4. The 'sans-taudis' is an *Armée du salut* neologism, indicative of the poetics of their weekly newspaper *En avant*! The term draws attention not only to the plight of the homeless, 'les sans logis' or 'sans abri', but also to the generally low standard of housing for the poor as a social category.
5. See, for example, G. Monnier, *Le Corbusier* (Mesnil-sur-l'Estrée: Editions la Manufacture, 1992), p. 47.
6. Foucault, 'Of Other Spaces', p. 34.
7. The Penal Colonies in French Guyana will be discussed later. They form an intrinsic part of the City of Refuge programme.
8. See Albert Flament in *En avant*, 15 June 1929 and *En avant*, 8 March 1930.
9. These factual, poetical and metaphorical topics will be explored in section one.
10. See *En avant*, 30 November 1929: 'The *Armée du salut* wants to injure the eyes of the French people with the Truth; it wants to thrust fists-full of light at them'. The *Armée du salut*'s activities between 1927–31, leading up to the barge/refuge project and their initial years, are the subject of Chapter 8, 'Water as Reality and Metaphor in *En avant*'.
11. See, for example, *En avant*, 8 March 1930.
12. *En avant*, 25 January 1930, Bibliothèque de la Société de l'Histoire du Protestantisme Français, Paris. During the summer time the boat moved upstream to Pecq where it 'metamorphosed' into a 'pleasure yacht' for young men who wouldn't otherwise be able to afford a holiday break; see *En avant*, 13 July 1929.
13. *En avant*, 5 July 1930, *En avant*, 26 July 1930, and *En avant* 10 August 1930.
14. *En avant*, 11 January 1930; and *En avant*, 25 January 1930.
15. *En avant*, 25 January 1930.
16. The poetic quality of many of the articles of *En avant* merits being better translated and better known.
17. *En avant*, 25 January 1930.
18. *En avant*, 25 January 1930.
19. The writings of *En avant* are keenly aware of just exactly how much the homeless are at the mercy of the elements.
20. *En avant*, 25 January 1930.
21. M. Foucault, *Le corps utopique, Les hétérotopies* (1966; Paris: Nouvelles éditions lignes, 2009), p. 24; M. Foucault, 'Des espaces autres', in M. Foucault, *Dits et écrits 1976–1988* (1967; Paris: Gallimard, 1984), p. 1574; M. Foucault, 'Of Other Spaces', trans. J. Miskowiec, *Diacritics*, 16:1 (Spring 1986), pp. 22–7, on p. 23.
22. Foucault, 'Des espaces autres', p. 1573; and Foucault, 'Of Other Spaces', p. 23. Foucault's later version of the 'heterotopia' piece 'Des espaces autres' in 1984 and 1986 scales

down the potential of utopian thinking and practice for bringing about radical societal change. The implications of, and maybe reasons for, this shift will be revisited in my conclusion.
23. See the discussion of the penal colonies (*les bagnes*) below.
24. F. Nietzsche, *Kritische Studienausgabe Nachgelassene Fragmente 1884-5* (Berlin: dtv/ de Gruyter, 1988), section 224.
25. *En avant*, 2 August 1930, Bibliothèque de la Société de l'Histoire du Protestantisme Français, Paris.
26. F. Engels, 'Socialism: Utopian and Scientific', in K. Marx and F. Engels, *Karl Marx and Frederick Engels: Selected Works*, 3 vols (Moscow: Progress Pubishers,1969–70), vol. 3, pp. 95–151.
27. The Salvation Army was founded by William Booth in the East End of London in 1878.
28. *En avant*, 1 June 1929 and *En avant*, 20 July 1929.
29. *En avant*, 16 November 1929.
30. *En avant*, 2 August 1930.
31. *En avant*, 2 August 1930, my italics.
32. *En avant*, 16 November 1929.
33. *En avant*, 15 February 1930.
34. *En avant*, 2 August 1930.
35. *En avant*, 26 July 1930.
36. *En avant*, 3 May 1930.
37. *En avant*, 16 February 1929.
38. *En avant*, 16 February 1929.
39. M. Foucault, *The Order of Things* (London: Tavistock, 1970), p. xv.
40. Foucault, *The Order of Things*, p. xv.
41. *En avant*, 28 January 1929. Le Corbusier uses exactly the same words in Le Corbusier, *Sur les quatre routes* (Paris: Denoël, 1941).
42. *En avant*, 3 August 1929.
43. See Figure 8.2 and note 7, above.
44. *En avant*, 20 July 1929.
45. For more on these colonies see D. Morgan, 'The "Floating Asylum", The *Armée du salut* and Le Corbusier: A Modernist Heterotopian/Utopian Project', *Utopian Studies*, 25:1 (2014), pp. 87–124.
46. See, for example, P. Hamp, 'Preface', in C. Péan, *Terre de bagne* (Paris: Editions Altis, 1933), pp. 7–24, on p. 11.
47. Hamp, 'Preface', in Péan, *Terre de bagne*, pp. 8–9.
48. Hamp, 'Preface', in Péan, *Terre de bagne*, p. 7.
49. *En avant*, 30 November 1929. 'There is no document of culture which is not at the same a document of barbarism'; W. Benjamin, 'On the Concept of History', in H. Eiland & M. Jennings (eds), *Walter Benjamin: Selected Writings*, 4 vols (Cambridge, MA: Belknap Press, 2003), vol. 4, pp. 389–400, on p. 392.
50. *En avant*, 30 November 1929; and Benjamin, 'On the Concept of History', p. 13.
51. *En avant*, 30 November 1929; and Benjamin, 'On the Concept of History', p. 16.
52. A. Peyron, 'Introduction', in C. Péan, *Terre de bagne* (Paris: Editions Altis, 1933), pp. 27–34, on pp. 28–30. See also J. Derrida on how unconditional ('disinterested') hospitality involves exposure and risk: J. Derrida and A. Dufourmantelle, *De l'hospitalité* (Paris: Calmann-Lévy, 1997).

53. See also *En avant*, 26 July 1930 for a vivid image of a wicked person tormented by nocturnal demons.
54. Peyron, 'Introduction', in Péan, *Terre de bagne*, p. 31.
55. *En avant*, 11 January 1930; and Hamp, 'Preface', in Péan, *Terre de bagne*, p. 15. 'Gueuserie' is another *Armée du salut* neologism.
56. Hamp, 'Preface', in Péan, *Terre de bagne*, p. 15; and Peyron, 'Introduction', in Péan, *Terre de bagne*, p. 32.
57. Peyron, 'Introduction', in Péan, *Terre de bagne*, p. 31.
58. *En avant*, 29 March 1930; and *En avant*, 12 July 1930.
59. See Joshua 20:1–9; Revelations 21:10–27; and Psalms 46:4; *Holy Bible: Authorised King James Version*.
60. P. J. Proudhon, *What is Property?*, ed. D. R. Kelly & B. Smith (Cambridge: Cambridge University Press, 2007), p. 69.
61. Proudhon, *What is Property?*, p. 216. Proudhon features in Le Corbusier, *Trois établissements humains* (1945; Paris: Les éditions de minuit, 1997). In the interesting section 'Conditions morales' written by Hyacinthe Dubreuil, a former worker and foreman in the car industry, there is an allusion to 'proudhonian philosophy' in relation to useful work and the joy of living; H. Dubreuil, 'Conditions morales', in Le Corbusier, *Trois établissements humains* (1945; Paris: Les éditions du Minuit, 1997), pp. 59–61. For a more sinister 'proudhonian' association, see G. Valois, 'La nouvelle étape du fascisme à la réussite par la pauvreté', *Le nouveau siècle*, 23 May 1927, La Fondation Le Corbusier, X1-4-80. Valois was a member of the extreme right-wing *Action Française* and the *Cercle Proudhon*. My thanks go to Tim Benton for a discussion about the ambiguity of Le Corbusier's personality.
62. *En avant*, 28 December 1929.
63. Le Corbusier, *Sur les quatre routes* (1941; Paris: Denoël, 1971), p. 15.
64. Evidently the City of Refuge and the *Unités d'habitation* resemble ocean liners.
65. Le Corbusier, *Vers une architecture* (Poitiers: Arthaud, 1977), pp. 70, 78.
66. Le Corbusier, *Sur les quatre routes*, p. 256.
67. Le Corbusier, *The Radiant City* (1933; London: Faber and Faber, 1967), p. 14.
68. For dystopian aspects of the 'asile flottant' and the City of Refuge itself as architectural projects, see D. Morgan Morgan, *'Globus Terraqueus*: Cosmopolitan Right, 'Fluid Geography' and *Commercium* in the Utopian Thinking of Immanuel Kant and Pierre-Joseph Proudhon', in *Law and the Utopian Imagination* (Stanford, CA: Stanford University Press, 2014).
69. See the image – with its caption 'awakening of cleanliness' – of the riots in Paris, 6th February 1934, of the combined forces of the extreme right which resulted in three hundred injured and fifteen deaths; Le Corbusier, *The Radiant City* (London: Faber & Faber, 1976), p. 23. See S. Richards, *Le Corbusier and the Concept of Self* (New Haven, CT, and London: Yale University Press, 2003), pp. 41–5, for an explanation of its antisemitic motivation and of *l'Action Française*. For the complimentary reference to Pétain see Le Corbusier, *The Radiant City*, p. 154.
70. Le Corbusier, *The Radiant City*, p. 154.
71. *En avant*, 11 January 1930.
72. 'Le faisceau' can be translated as 'The Bundle' or 'The Cluster', or even 'The Network'.
73. Valois, 'La nouvelle étape du fascisme à la réussite par la pauvreté'.
74. Valois, 'La nouvelle étape du fascisme à la réussite par la pauvreté'.

75. Of course with utopias one should always ask the question: 'whose utopia?'. Fascists would not consider their worldview to be 'dystopian' at all...
76. M. Foucault, 'Des espaces autres', in M. Foucault, *Dits et écrits 1976–1988* (1967; Paris: Gallimard, 1984), pp. 1571–1581, on p. 1575; and M. Foucault, 'Of Other Spaces', trans. J. Miskowiec, *Diacritics*, 16:1 (Spring 1986), pp. 22–7, on p. 24.
77. M. Foucault, *Le corps utopique, Les hétérotopies* (1966; Paris: Nouvelles éditions lignes, 2009), p. 24; Foucault, 'Des espaces autres', p. 1574; and Foucault, 'Of Other Spaces', p. 24.
78. Defert, in M. Foucault, *Les mots et les choses* (Paris: Gallimard, 2009).
79. M. Gauchet and G. Swain, *La pratique de l'esprit humain* (Paris: Gallimard, 2007), p. 202.
80. M. Foucault, *Madness and Civilisation*, trans. R. Howard (London: Tavistock, 1982), p. 38.
81. M. Abensour, *Lettre d'un 'révoltiste' à Marcel Gauchet converti à la 'politique normale'* (Paris: Sens &Tonka, 2008), p. 18.
82. Abensour, *Lettre d'un 'révoltiste' à Marcel Gauchet*, p. 19.

9 Zooheterotopias

1. B. Heuvelmans, 'Preface', *On the Track of Unknown Animals*, trans. R. Garnett, 3rd edn (London: Paladin, 1965), cited in 'Definition of cryptozoology', at http://cryptozoo.monstrous.com/definition_of_cryptozoology.htm [accessed 20 July 2014].
2. Heuvelmans, *On the Track of Unknown Animals*, p. 49.
3. I. T. Sanderson, quoted in B. Regal, *Searching for Sasquatch: Crackpots, Eggheads and Cryptozoology* (New York: Palgrave MacMillan, 2011), p. 110.
4. Heuvelmans, *Unknown Animals*, p. 1.
5. Regal, *Searching for Sasquatch*.
6. Regal, *Searching for Sasquatch*, p. 23.
7. See Regal, *Searching for Sasquatch*, pp. 12–13 for a brief cryptozoological account of medieval bestiaries and other early sources.
8. Heuvelmans, *Unknown Animals*, p. 20.
9. Heuvelmans, *Unknown Animals*.
10. R. Luckhurst, *Alien* (Basingstoke: Palgrave MacMillan, 2014).
11. Regal, *Searching for Sasquatch*, p. 1.
12. M. Foucault, 'Of Other Spaces', trans. J. Miskowiec, *Diacritics*, 16:1 (Spring 1986), pp. 22–7, on p. 24.
13. C. Taylor, 'Foucault and Critical Animal Studies: Genealogies of Agricultural Power', *Philosophy Compass*, 8:6 (2009), pp. 539–51, on p. 539.
14. S. Rinfret, 'Controlling Animals: Power, Foucault and Species Management', *Society and Natural Resources*, 22:6 (2009), pp. 571–8, on p. 572.
15. J. Conrad, *Heart of Darkness* (London: Penguin Classics, 1973), p. 12.
16. Heuvelmans, *Unknown Animals*, p. 160; emphasis added.
17. Heuvelmans, *Unknown Animals*, p. 17.
18. A. Conan Doyle, *The Lost World* (Oxford: World's Classics, 1995), p. 12.
19. H. Rider Haggard, *The Days of My Life*, 2 vols.
20. H. Rider Haggard, 'Elephant Smashing and Lion Shooting', *African Review*, 3 (June 1894), pp. 762–3, on p. 762.
21. The status of imperial capital in Haggard's writing and in adventure fiction more

generally is a complex topic. Readers of *King Solomon's Mines*, to give perhaps the most famous example, will unavoidably detect an economic imperative that has been connected to the material exploitation of South Africa by numerous critics. See, for example, L. Chrisman, *Rereading the Imperial Romance: British Imperialism and South African Resistance in Haggard, Schreiner and Plaatje* (Oxford: Oxford University Press, 2000); and A. McLintock, *Imperial Leather: Race, Gender and Sexuality in the Colonial Conquest* (London: Routledge, 1995).

22. The year 2012, however, saw the launch of a new academic journal, *Journal of Cryptozoology*.
23. M. Foucault, 'Of Other Spaces', trans. J. Miskowiec, *Diacritics*, 16:1 (Spring 1986), pp. 22–7, on p. 26.
24. Foucault, 'Of Other Spaces', p. 26.
25. Foucault, 'Of Other Spaces', p. 27.
26. Lovelace's novel, D. W. Lovelace, *King Kong* (Nevada City, CA: Underwood Books, 2005), though relatively obscure next to the iconic movie it accompanies, does in many scenes provide a richer and more provocative engagement than the film with the ecological and economic discourses I am interested in here. While the visual spectacle that cinema demands sees the movie reduced to a succession of chase scenes with minimal dialogue (unless the Ann Darrow's repeated screams be counted as such), Lovelace's novel gives greater insight into the ideological discourses at work in the formation of *King Kong*.
27. D. W. Lovelace, *King Kong* (Nevada City, CA: Underwood Books, 2005), p. 31.
28. We might also think of the development of science fiction as a parallel movement to the emergence of cryptozoology. Indeed, there are branches of cryptozoology that are very much concerned with alien life, or what we might call galactic biodiversity.
29. Lovelace, *King Kong*, p. 74.
30. Lovelace, *King Kong*, p. 59.
31. Lovelace, *King Kong*, pp. 75–6.
32. Lovelace, *King Kong*, p. 59.
33. A. Conan Doyle, *The Lost World* (Oxford: World's Classics, 1995), p. 127.
34. This observation has been made in numerous critical studies of imperial fiction, particularly in the analysis of Rider Haggard's writing. See, for example, A. McLintock, *Imperial Leather: Race, Gender and Sexuality in the Colonial Conquest* (London: Routledge, 1995).
35. M. Foucault, 'Of Other Spaces', trans. J. Miskowiec, *Diacritics*, 16:1 (Spring 1986), pp. 22–7, on p. 26.
36. Conan Doyle, *The Lost World*, p. 8.
37. Conan Doyle, *The Lost World*, p. 57.
38. Conan Doyle, *The Lost World*, p. 72.
39. Conan Doyle, *The Lost World*, p. 73.
40. Conan Doyle, *The Lost World*, p. 112.
41. Conan Doyle, *The Lost World*, p. 112.
42. Conan Doyle, *The Lost World*, p. 114.
43. Interestingly, Kong shares the human disgust for reptilian life. Overcoming a dinosaur after a breathless struggle, the outsized gorilla 'shivered, so full he was of the loathing his species has had of reptilian things since the dawn of time', Lovelace, *King Kong*, p. 111.
44. M. Foucault, 'Of Other Spaces', trans. J. Miskowiec, *Diacritics*, 16:1 (Spring 1986), pp.

210   *Notes to pages 159–65*

   22–7, on p. 26.
45. A. Conan Doyle, *The Lost World* (Oxford: World's Classics, 1995), p. 75.
46. Conan Doyle, *The Lost World*, p. 76.
47. Conan Doyle, *The Lost World*, p. 77.
48. Foucault, 'Of Other Spaces', p. 26.
49. D. W. Lovelace, *King Kong* (Nevada City, CA: Underwood Books, 2005), p. 31–2.
50. Lovelace, *King Kong*, p. 31.
51. Lovelace, *King Kong*, p. 71.
52. Lovelace, *King Kong*, p. 71.
53. *The Lost World* is interestingly resistant to the convergence of colonial exploration and heterosexual seduction. Despite proving himself in the testing ground of the South American frontier, Malone does not in the end get the girl, whose affections are won instead by a 'solicitor's clerk', Conan Doyle, *The Lost World*, p. 204.
54. The two most widely cited sources on the relationships between conservation and imperialism are R. H. Grove, *Green Imperialism* (Cambridge: Cambridge University Press, 1995); and A. Crosby, *Ecological Imperialism* (Cambridge: Cambridge University Press, 1986).
55. M. Foucault, 'Of Other Spaces', trans. J. Miskowiec, *Diacritics*, 16:1 (Spring 1986), pp. 22–7, on p. 27.
56. A. Conan Doyle, *The Lost World* (Oxford: World's Classics, 1995), p. 5.
57. R. G. Forman, 'Room for Romance: Playing with Adventure in Arthur Conan Doyle's *The Lost World*', *Genre*, 43 (2010), pp. 27–59, on p. 49.
58. D. W. Lovelace, *King Kong* (Nevada City, CA: Underwood Books, 2005), p. 2.
59. Lovelace, *King Kong*, p. 6.
60. Lovelace, *King Kong*, p. 137.
61. See Regal, *Searching for Sasquatch* on how real-life cryptozoologists were often accused of hoaxing and huckstering for financial gain.
62. B. Creed, *Darwin's Screens: Evolutionary Aesthetics, Time and Sexual Display in Cinema* (Melbourne: Melbourne University Press, 2009), p. 11.
63. Lovelace, *King Kong*, p. 134.
64. Lovelace, *King Kong*, p. 140.
65. The reading of Kong as Christ is one of several allegorical interpretations that Noel Carroll lists in N. Carroll, '*King Kong*: Ape and Essence', in B. K. Grant and C. Sharrett (eds), *Planks of Reason: Essays on the Horror Film* (Oxford: Scarecrow Press, 2004), pp. 15–16.
66. Conan Doyle, *The Lost World*, p. 202.
67. Forman, 'Room for Romance', p. 54.
68. Foucault, 'Of Other Spaces', p. 27.

10 Soft Machines

1. The research for this chapter was carried out during, and made possible by, a postdoctoral fellowship at the Institute for Advanced Studies in the Humanities at the University of Edinburgh. M. Foucault, 'Of Other Spaces', trans. J. Miskowiec, *Diacritics*, 16:1 (Spring 1986), pp. 22–7, on p. 24.
2. Foucault. 'Of Other Spaces', p. 24.
3. See F. Jameson, *Archaeologies of the Future: The Desire Called Utopia and Other Science Fictions* (London and New York: Verso, 2007), p. xii.

Notes to pages 165–71   211

4. P. E. Wegner, 'Horizons, Figures, Machines: The Dialectic of Utopia in the Works of Fredric Jameson', *Utopian Studies*, 9:2 (1998), pp. 58–77, on p. 58.
5. Jameson, *Archaeologies of the Future*, p. xii.
6. Jameson, *Archaeologies of the Future*, p. xiii.
7. Jameson, *Archaeologies of the Future*, p. xii.
8. Jameson, *Archaeologies of the Future*, pp. 1, 3.
9. Jameson, *Archaeologies of the Future*, pp. 211–12, 231–2.
10. Foucault, 'Of Other Spaces', p. 24.
11. Foucault, 'Of Other Spaces', p. 24.
12. Foucault, 'Of Other Spaces', p. 22.
13. Foucault, 'Of Other Spaces', pp. 22, 24.
14. Foucault, 'Of Other Spaces', p. 24.
15. Foucault, 'Of Other Spaces', p. 26.
16. Foucault, 'Of Other Spaces', p. 26.
17. J. Derrida, *Archive Fever: A Freudian Impression*, trans. E. Prenowitz (Chicago, IL: University of Chicago Press, 1998), pp. 11–12.
18. Derrida, *Archive Fever*, pp. 11–12.
19. See J. Derrida, 'No Apocalypse, Not Now (Full Speed Ahead, Seven Missiles, Seven Missives), *Diacritics*, 14:2 (Summer 1984), pp. 20–31.
20. Foucault, 'Of Other Spaces', p. 22.
21. Foucault, 'Of Other Spaces', p. 22.
22. Whitelee Wind Farm, at http://www.whiteleewindfarm.co.uk/ [accessed 13 February 2014].
23. Foucault, 'Of Other Spaces', p. 26.
24. Whitelee Wind Farm, at http://www.whiteleewindfarm.co.uk/ [accessed 13 February 2014].
25. Jameson, *Archaeologies of the Future*, p. 3.
26. Foucault, 'Of Other Spaces', p. 24.
27. Foucault, 'Of Other Spaces', p. 27.
28. The verb 'cross-hatches' comes from C. Miéville, *The City and The City* (London: Pan Books, 2009).
29. M. Foucault, 'Of Other Spaces', trans. J. Miskowiec, *Diacritics*, 16:1 (Spring 1986), pp. 22–7, on p. 24.
30. Foucault, 'Of Other Spaces', p. 24.
31. Foucault, 'Of Other Spaces', p. 24.
32. For this discussion of the maze, I made reference to K. Veel, 'The Irreducibility of Space: Labyrinths, Cities, Cyberspace', *Diacritics*, 33:3–4, New Coordinates: Spatial Mapping, National Trajectories (Autumn–Winter 2003), pp. 151–72.
33. J. J. Abrams (dir.), *Mission Impossible III* (Paramount, 2006).
34. C. Nolan (dir.), *Inception* (Warner Brothers, 2010).
35. D. Dickson, *Alternative Technology and the Politics of Technical Change* (Glasgow: Fortuna, 1974), p. 38.
36. Foucault, 'Of Other Spaces', p. 24.
37. F. Jameson, *Archaeologies of the Future: The Desire Called Utopia and Other Science Fictions* (London and New York: Verso, 2007), p. 213.
38. M. McLuhan, *Understanding Media: The Extensions of Man* (London and New York: Routledge, 2010), p. 11.
39. M. Zardini, '(Against) the Greenwashing of Architecture', *New Geographies*, 2 (Novem-

ber 2009), pp. 139–147, on p. 146.
40. Foucault, 'Of Other Spaces', p. 24.
41. L. Mumford, *Technics and Civilisation* (Chicago, IL: University of Chicago Press, 2010), p. 163.
42. D. DeLillo, *White Noise* (London: Picador, 1999), p. 301.
43. DeLillo, *White Noise*, p. 4.
44. McLuhan, *Understanding Media*, p. 45.
45. McLuhan, *Understanding Media*, pp. 46, 61.
46. McLuhan, *Understanding Media*, pp. 12, 52.
47. F. Léger, 'The Machine Aesthetic: The Manufactured Object, the Artisan, and the Artist', in E. J. Fry (ed.), *Functions of Painting*, trans. A. Anderson (London: Thames & Hudson, 1973), pp. 52–61, on p. 54.
48. R. Barthes, *The Eiffel Tower and Other Mythologies*, trans. R. Howard (New York: Hill & Wang, 1979), p. 4.
49. Barthes, *The Eiffel Tower*, pp. 4, 7.
50. R. Barthes, *The Eiffel Tower and Other Mythologies*, trans. R. Howard (New York: Hill & Wang, 1979), p. 3.
51. Barthes, *The Eiffel Tower*, p. 5.
52. H. G. Wells, *The War of the Worlds* (London: J. M. Dent, 2003).
53. See, for example, D. Suvin, 'On the Poetics of the Science Fiction Genre', *College English*, 34:3 (December 1972), pp. 372–83.
54. Virilio comments on the bunker in these terms in P. Virilio, *Bunker Archeology*, trans. G. Collins (New York: Princeton Architectural Press, 2008), p. 12.
55. See A. Parr, *Hijacking Sustainability* (Cambridge, MA: Massachusetts Institute of Technology Press, 2009). The discussion to follow is heavily indebted to his study. Thanks to Lisa Moffitt for bringing this to my attention.
56. Parr, *Hijacking Sustainability*.
57. F. T. Marinetti, 'Geometric and Mechanical Splendour and the Numerical Sensibility', in U. Apollonio (ed.), *Futurist Manifestos* (London: Thames & Hudson, 1973), p. 154.
58. F. Léger, 'The Machine Aesthetic: The Manufactured Object, the Artisan, and the Artist', in E. J. Fry (ed.), *Functions of Painting*, trans. A. Anderson (London: Thames & Hudson, 1973), pp. 52–61, on p. 62.
59. Norman Bel Geddes's term that replaces the physical terms, in streamlining design, 'eddies' and 'partial vacua' that had to be elimited in order to achieve 'perfection'. See C. Cogdell, *Eugenic Design: Streamiling America in the 1930s* (Philadelphia, PA: University of Pennsylvania Press, 2004), p. 51.
60. See Cogdell, *Eugenic Design*, chapter 2, pp. 33–83 on the correlations between eugenics and streamline design.
61. E. Burke, *A Philosophical Enquiry Into the Origin of Our Ideas of the Sublime and Beautiful* (London: Basil Blackwell, 1987), p. 80.
62. Lèger, 'The Machine Aesthetic', p. 60.
63. S. Freud, 'Fetishism', in J. Strachey (ed.), *Standard Edition of the Complete Psychological Works of Sigmund Freud*, 24 vols(London: Vintage, 1974), vol. 21, p. 351.
64. Freud, 'Fetishism', p. 351.
65. Freud, 'Fetishism', p. 351.
66. Freud, 'Fetishism', pp. 352–3.
67. Freud, 'Fetishism', p. 354.
68. Freud, 'Fetishism', p. 353.

69. T. Pynchon, *Gravity's Rainbow* (London: Vintage, 2000), p. 393.
70. Pynchon, *Gravity's Rainbow*, p. 395.
71. Pynchon, *Gravity's Rainbow*, p. 487.
72. Pynchon, *Gravity's Rainbow*, p. 487–8.
73. Pynchon, *Gravity's Rainbow*, p. 249.
74. Pynchon, *Gravity's Rainbow*, pp. 724, 750.
75. V. E. Yarsley and E. G. Couzens, *Plastics* (Harmondsworth and New York: Penguin Books, 1941), p. 16.
76. Yarsley and Couzens, *Plastics*, p. 11.
77. Yarsley and Couzens, *Plastics*, pp. 121, 154.
78. Yarsley and Couzens, *Plastics*, pp.120, 134.
79. It is Judith Brown who performs such an analysis on the usage of cellophane in theatre productions, films, packaging, in her article J. Brown, 'Cellophane Glamour', *Modernism, Modernity*, 15:4 (November 2008), pp. 605–26.
80. J. L. Meikle, 'Materia Nova: Plastics and Design in the US, 1925–1935', in S. T. I. Mossman and P. J. T. Morris (eds), *The Development of Plastics* (London: Royal Society of Chemistry, 1994), pp. 38–53, on p. 45, cited in Brown,'Cellophane Glamour', p. 608.
81. J. L. Meikle, *American Plastic: A Cultural History* (New Brunswick, NJ: Rutgers University Press, 1997), pp. 1–2.
82. On Rocket Societies see F. H. Winter, *Prelude to the Space Age: The Rocket Societies, 1924–1940* (Washington, D.C.: Smithsonian Press, 1983).
83. P. T. Frankl, *Form and Re-Form: A Practical Handbook of Modern Interiors* (New York: Harper & Brothers, 1930), pp. 5, 163.
84. P. T. Frankl, *Machine-Made Leisure* (New York: Harper & Brothers, 1932), p. 12.
85. J. Gloag, *Plastics and Industrial Design* (London: George Allen & Unwin Ltd, 1945), p. 17.
86. Gloag, *Plastics and Industrial Design*, p. 53.
87. Gloag, *Plastics and Industrial Design*, pp. 45, 53.
88. Yarsley & Couzens, *Plastics*, p. 10.
89. Yarsley & Couzens, *Plastics*, p. 12.
90. R. Barthes, *Mythologies*, trans. A. Lavers (London: Vintage, 1993), p. 97.
91. D. F. Wallace, *Infinite Jest* (London: Abacus, 2009), p. 381. I have written about*Infinite Jest*and plastics elsewhere; see F. Collignon, 'USA Murated Nation, or, The Sublime Sphereology of Security Culture',*The Journal of American Studies*, Vol. 48, No. 3 (August 2014), pp. 1–25.
92. Meikle, *American Plastic*, p. 195.
93. Meikle, *American Plastic*, p. 140. On Du Pont, see also J. Pavitt, *Fear and Fashion in the Cold War* (London: V&A Publishing, 2008); S. Handley, *Nylon: The Story of a Fashion Revolution* (Baltimore, MD: Johns Hopkins University Press, 1999) and P. A. Ndiaye, *Nylon and Bombs: Du Pont and the March of Modern America*, trans. E. Forster (Baltimore, MD: Johns Hopkins University Press, 2006).
94. Meikle, *American Plastic*, p. 128.
95. Meikle, *American Plastic*, p. 207.
96. J. Ellul, *The Technological Society*, trans. J. Wilkinson (London: Jonathan Cape, 1965), p. 85.
97. K. Hetherington, *The Badlands of Modernity: Heterotopia and Social Ordering* (London and New York: Routledge, 1997), p. vii.

INDEX

Abensour, Miguel, 147
Abrams, Jeffrey Jacob
   Mission Impossible III, 174, 177
Ackroyd, Peter, 7, 32, 37–9, 41, 44
Africa, 6, 9, 10, 56, 92, 98–9, 101–2, 106–8, 117, 130
Agamben, Giorgio, 85
AIDS, 52
Alexander, Elizabeth, 86
Althusser, Louis, 14
Arabs, 98–9, 103–4, 109
Arlesheim, Switzerland, 86
Armée du salut (French Salvation Army), 11, 127, 132, 136, 145
Ascoli, Graziadio Isaia, 26

Bachelard, Gaston, 14
Ballard, James Graham, 30
Barthes, Roland, 12, 173, 180
   Mythologies, 172
Bartoli, Matteo Giulio, 26
Bateson, Gregory, 37
Battle Of The Nile, 118
Baudrillard, Jean, 31, 75
Bay of Naples, 117
Beckett, Samuel
   Endgame, 39
Benjamin, Walter, 111, 119–20, 124
Bergson, Henri, 19
Berlin, 125, 166, 169
Bhambra, Gurminder
   Rethinking Modernity, 126
biomythography, 88
Black Atlantic, 116–17, 121
Blanchot, Maurice, 6, 50
Blixen, Karen, 9–10, 98–101, 107
   Out of Africa, 10

Bloch, Marc, 14
boat, 11, 37, 55, 59, 79, 118, 127, 129, 130, 134, 143, 146, 168
Borges, Jorge Luis, 2
Bourbon Kingdom, 20
Braudel, Fernand, 14
Britain, 44, 58–9, 66–8, 71, 73, 90, 92
Brixton, 52–4, 59, 60
Bucharin, 20
Bunhill Fields, 41
Burgio, Alberto, 19
Burke, Edmund, 176
   A Philosophical Enquiry Into The Origins Of Our Ideas Of The Sublime And Beautiful, 176
Burke, Lucy, 90
Burroughs, Edgar Rice, 11, 151, 155

Cairo, 122
Calicut, 100
camp, 39, 68, 71, 76, 78, 85
cancer, 9, 81, 82, 86–8
Canguilhem, Georges, 14
capital, 5–7, 10–12, 30–5, 38, 45–7, 111–12, 114–15, 127, 143, 151–3, 155, 160, 162, 164
Caribbean, 92, 116–18, 121
Carter, Angela, 30
Central Africa, 101
Chambers, Iain, 25
Cheyenne River Sioux Reservation, 83
Chikwava, Brian, 8, 51, 58–60, 64
   Harare North, 8, 50, 55, 63
City of Refuge, 129, 134, 136, 139, 140, 143, 146
cleanliness, 72–3, 175
Cold War, 12, 174

– 215 –

Columbus, Christopher, 119
Conan Doyle, Arthur, 150, 154, 156, 159
   The Lost World, 11, 161, 164
Congo, 101–2
Conrad, Joseph, 9, 95
   'A Smile of Fortune', 9, 95, 98
Creed, Barbara, 162
Croce, Benedetto, 19, 21, 26
Crown Derby, 78
cryptozoological, 12
cryptozoology, 11, 149, 150–5, 159, 164
cybernetics, 34, 37

Darwin, Charles, 159
De Cauter, Lieven, 2, 6
Defoe, Daniel, 34, 41
Dehaene, Michiel, 2, 6, 83
Deleuze, Gilles, 16 *see also* Deleuze, Gilles and Felix Guattari
   Anto Oedipus, 16
Deleuze, Gilles and Felix Guattari, 29, 65, 68–70, 74–5, 79, 80, 121, 124
   A Thousand Plateaus, 16
dementia, 9, 81, 89–92
Derrida, Jacques, 52, 166
   Babel, 52
deterritorialized, 70–1
deviation, 83
Dick, Philip K., 30
Dietz, Albert, 180
disease, 81–2, 84, 86, 90, 145
Downes, Kerry, 43

Eaglesham Moor, 168
Egypt, 118
Eiffel Tower, 172–3
Ellul, Jacques, 181
English Peasants' Revolt (1381), 40
énoncés, 33
environment, 12
environmental, 3, 11, 15, 17, 25, 27, 29, 38, 151–4, 157, 159, 171, 174
Eshun, Kodwo and Steve Goodman, 121
etiology, 16
Europe, 14, 22–3, 31, 68, 86, 107, 114, 118–20, 123, 125, 136, 141
Euston station, 59

Faubian, James, 1
Febvre, Lucien, 14
feminist geography, 17
First World War, 127
Fischer, Robert, 170
Fordism, 7, 23
Foucault, Michel, 1–9, 11–22, 24–5, 27–8, 111
   Discipline and Punish, 14
   Le langage de l'espace, 14
   Les Hétérotopies, 13
   Madness and Civilization, 14
   Panopticum, 14
   The Birth of the Clinic, 14
   The Order of Things, 2
   'Of Other Spaces', 1, 3, 4, 50–1, 61, 68, 79, 81–3, 89, 92, 95, 146, 155, 164–5
   terrains vagues, 70
   'The Eye of Power', 3
Fourier, Charles, 139
Frankl, Paul, 179
Fraser, Nancy, 5
French Guyana, 143
French, Patrick, 103
French Revolution, 118
futurists, 175

Galapagos, 159
Gauchet, Marcel, 147
genealogy, 7, 15–16, 25
geohistory, 49, 52
German, 107, 153, 176, 179
Glasgow, 167
Gloag, John, 179
Gore, Al, 12, 174
Gramsci, Antonio, 6, 7, 15, 18, 19–26
   Prison Notebooks, 18–19
   Quaderni del carcere, 19–20
Grant, Linda, 89–91
   Remind Me Who I Am Again, 9
Green Bombers, 51, 53, 57
Guha, Ranajit, 21, 120
Gypsy Travellers, 8, 9, 66–77

Hackney, 8, 32, 45
Haggard, Henry Rider, 154–5
Haggerston Park, 38
Haiti, 117, 125
Harare, 8, 50–1, 55, 58, 60, 63–4

Index

Harvey, David, 27–8, 33
Hawksmoor, Nicholas, 30, 41–4
Hegel, Georg Wilhelm Friedrich, 26–7
hegemony, 3, 6–7, 15, 21–3, 26, 28, 47, 75, 116–18
Henriques, Julian, 124
heritage, 29, 30, 32, 33, 35–6, 39, 40–2, 44, 46–7
heterochronic, 50, 58–9, 63
Hetherington, Kevin, 93
Heuvelmans, Bernard, 149–51, 153–5
Heynen, Hilda, 1, 49
history, 16
HIV, 52
Holmes, Sherlock, 53
hospital, 39, 58, 89, 91
Huyssen, Andreas, 30

illness, 9, 81–2, 85, 87–9, 91, 93
immigrant, 59, 62, 84, 92
Indian Ocean, 95, 97–8, 100–1, 103
inside-outside, 73
Israel, 118

Jameson, Fredric, 12, 165, 167
Jim-Crow, 88
Johnson, Linton Kwesi, 123
Johnson, Peter, 1–2, 6, 8

Kant, Immanuel, 26–8, 187
Klinik, Lukas, 86, 87, 88
knowledge, 7, 10, 13–17, 22, 24, 26–8, 32, 34, 38, 44, 49, 64, 99, 100, 103, 105, 112–16, 123–6, 150, 154

Lakota, 82–3
Latour, Bruno, 3, 5, 29
Le Corbusier, 10–11, 127, 136, 145
    The Radiant City, 145
Lefebvre, Henri, 65, 68, 75, 113
Lèger, Fernand, 173, 175–6
Léothaud, Gilles and Bernard Lortat-Jacob, 123
liminality, 58
Livingstone, David, 99
London, 6–8, 30–3, 37–46, 50–5, 58–9, 61–4, 90, 136, 163
London Metropolitan Police Force Territorial Support Group (TSG), 40
London Review of Books, 12

Lorde, Audre, 86–9
    'A Burst of Light: Living with Cancer', 9
    The Cancer Journals, 9
Louise Catherine (coal barge), 10–11
Louise Catherine (painter), 11
Lukacs, György, 20
Lyotard, Jean-François, 14

MacLuhan, Marshall, 12
Malindi, 100
map, 30, 38, 112–13, 115, 123, 154–5
Marinetti, Filippo Tommaso, 175
maritime criticism, 119
Marker, Chris, 119
Marshall, Henrietta
    Our Island Story: A History Of England For Boys And Girls, 38
Marx, Karl, 18–19, 26
Mauritius, 9, 95–8
Mediterranean, 6, 10, 117–18, 120–3
Meillet, Paul Jules Antoine, 26
Mezzadra, Sandro, 114
Middle East, 118
migrant, 10, 50, 52, 58–9, 63, 69, 118–19
mimesis, 17
mobile home, 73, 76, 78–9
Mombasa, 103, 106, 108
Moorcock, Michael, 30
Morton, Timothy
    Ecology Without Nature, 174
Mother Africa, 56
Mugabe, Robert, 51–3
Mumford, Lewis, 172
music, 10, 88, 109, 122–4
Muslim, 100–1, 103, 122

Naipaul, Shiva, 9–10
    North of South, 10
Naipaul, Vidiadhar Surajprasad, 9, 101–3, 105–8
    A Bend in the River, 10
Napoleon, 118
Narcissus myth, 172
Native American, 83–6
Nazi, 149, 177
Neilson, Brett, 114
Nelson, 117
Nettl, Bruno, 122

New York, 66, 86, 90, 107, 163, 179
New Zealand, 149, 153
Nietzsche, Friedrich Wilhelm, 134
Nile, 99
Nolan, Christopher, 12
   Inception, 12, 170
nomad, 66, 68–70, 139
Northampton, 39

objectification, 24
occult geometry, 45
Ochsner, Barbara, 89
Olympic Games, 6–7
Omani, 99
Orientalism, 99
otherness, 9

Palestine, 118
Paris, 11, 45, 117, 127, 129, 130, 132, 134, 136, 146, 149, 176, 179
Péan, Charles, 143
Peyron, Albin, 136, 139, 143
Philo, Chris, 16
philology, 24, 25
placelessness, 50, 57–8, 61–4
Pohl, Frederik and Cyril Kornbluth, 170
Port Louis, 95, 97–8, 109
Portuguese, 100, 103
postcolonial, 9, 17, 45, 51, 55, 58, 63, 102–3, 113
postcoloniality, 8, 49, 58
postlapsarian, 51
power, 16
psychogeography, 30, 47
Public Order Act, 67
Pynchon, Thomas, 180

racism, 49, 85, 87, 97
Raffestin, Claude, 13
Ranke, Leopold Von, 20
Raygun, Ronnie, 174
Read, Jason, 5
refugee, 3, 85
regeneration, 29, 33, 39, 41–2, 47
renewable, 167, 171–3, 181
reverse archaeology, 36, 44
Risorgimento, 20–1
Roads Scotland Act, 67

Rocket Societies, 179
Russian revolution, 22

Said, Edward, 10, 21, 23, 25, 28, 122
Saint Domingue, 117
Saldanha, Arun, 2
Sanderson, Ivan T., 149
San Francisco, 179
security, 7, 30, 33–5, 38, 43, 46, 53, 106, 151, 174
sedentarism, 68, 72, 75, 76
sedentarization, 65
Sekula, Allan, 116
Selvon, Sam, 58, 62–3
   Lonely Londoners, 8, 50, 55
Serres, Michel, 18, 29
Sheeler, Charles, 175
ship, 9, 49, 76, 79, 95, 98, 116, 174
Sinclair, Iain, 7–8, 30–47
Sioux Falls, 84
Sioux Sanitorium, 83,–5
Skull Island, 156, 160–3
slavery, 49, 104
Soja, Edward, 1, 49, 52, 73, 76
Sorel, Georges Eugène, 19
south, 20–1, 123, 164
South America, 154, 157, 162
South Dakota, 83
Southern Question, 7, 20, 22–4
space, 3, 7–8, 13, 16
spatialization, 8–9, 14, 65
Spivak, Gayatri Chakravorty, 21
Steiner, Rudolf, 87
St Katharine's Dock, 40
St Pierre, Mark, 83
Strategic Defence Initiative (SDI), 174
subaltern studies, 21, 25, 28
subjection, 24–5
Swaheli, 98, 100
Swain, Gladys, 147
Swan, Madonna, 82, 83–6, 88
   Madonna Swan: A Lakota Woman's Story, 9
Switzerland, 86
syntax, 53, 93, 113

Tanner, Laura, 81
techno-fetishism, 172
Thames, 7, 37–9, 44

Thatcher, Margaret, 30
Theroux, Paul, 103
Thrift, Nigel, 3
time, 4, 8
time-space compression, 30, 33
Topinka, Robert J., 1
trialectic of spatiality, 52
tropological space, 54, 63
tuberculosis, 9, 81, 83, 85
turbine technology, 172
Tyler, Imogen, 68, 73, 74
   Revolting Subjects, 74

Uganda, 101, 103
Umm Kalthūm, 10, 122
United States
   Public Health Service, 83
urban, 2, 7, 11, 16, 29, 30, 31, 40, 45, 47, 68, 107, 122, 130, 157, 162, 167

Valois, Georges, 146
Venus, 170
Victorian Christianity, 127

Wallace, Edgar, 155
   King Kong, 11, 151

Walsh, Mikey, 8, 66, 76
   Gypsy Boy, 8
Warner, Marina, 12
   London Review of Books, 12
Warren Woods Caravan Park, 70
Waswahili, 99, 103, 105
Waterloo, 59
Wegner, Phillip E., 165
west, 38, 44, 164
West Sussex, 76
Whitelee Wind Farm, 167, 170
Woolf, Virginia
   'On Being Ill', 82, 91

Yarsley, Victor Emmanuel and Edward Gordon Couzens, 178
yeti, 149, 154

Zaire, 107
Zambia, 101, 105, 107
Zanzibar, 99, 102
Zillhardt, Madeleine, 11, 127, 130
Zimbabwe, 50–1, 53–4, 58, 60, 62
Žižek, Slavoj, 171
zooheterotopias, 151, 159
Zürich, 88

Printed in Great Britain
by Amazon